Banning Chemical Weapons provides a brief and readable outline of the scientific topics that underlie efforts to ban chemical weapons. The aim of this volume is to give the general reader an appreciation of how scientific material must be used in negotiations to achieve an enforceable worldwide ban on chemical weapons. It also offers an example of how other disarmament measures must include similar technical considerations. The book progresses from the history of chemical warfare, through the technology of offence and defence, to the technical problems facing negotiators of a disarmament treaty.

Banning chemical weapons

Banning chemical weapons

The scientific background

Hugh D. Crone

Senior Principal Research Scientist
Materials Research Laboratory, Melbourne, Australia

Published by the Press Syndicate of the University of Cambridge
The Pitt Building, Trumpington Street, Cambridge CB2 1RP
40 West 20th Street, New York, NY 10011-4211, USA
10 Stamford Road, Oakleigh, Victoria 3166, Australia

First published 1992

Printed in Canada

Library of Congress Cataloging-in-Publication Data
Crone, Hugh D.
Banning chemical weapons: the scientific background / Hugh D. Crone.
p.   cm.
Includes index.
ISBN 0-521-41699-X (hardback). – ISBN 0-521-42711-8 (pbk.)
1. Chemical warfare.   2. Chemical arms control.   I. Title.
UG447.C76   1992
358'.34 – dc20   92-8882
                                  CIP

A catalog record for this book is available from the British Library

ISBN 0-521-41699-X hardback
ISBN 0-521-42711-8 paperback

Contents

v

Preface

Peace is the most elusive of quarries; it draws many into the marshes and swamps of policy, statesmanship, national self-interest, factional jealousies, true fellowship, grand posturings, honest endeavour, a mass of good intentions bogged in human frailty. If progress is to be made, it must come from a slow process of identifying, isolating and removing the causes of war. This work must involve scientists in the origins of war (briefly considered in Chapter 2) and in the technology of certain types of warfare, such as chemical warfare. This book is purely technical, yet I believe that an understanding of technical issues is necessary for nonscientists interested in achieving a ban on chemical warfare that will really work, that is, will be effective where the Geneva Protocol of 1925 patently was not.

As I write this in early 1992, there is optimism that a Chemical Weapons Convention could be signed later this year. However, this optimism is tempered by uncertainties such as differences in attitude between north and south, the future of the former Soviet republics, and the confused situation in the Middle East. Therefore, you may be reading this volume either as a statement of technical problems to be solved quickly or as a presentation of technical reasons why we failed yet again to secure a convention. The signing of such an agreement will be the first step of a ten-year effort to set in place and activate a technical organisation to monitor the convention in perpetuity, again an activity dependent on knowledge of the matters described in this book.

I have not set out to compete with the many books now available that give the history of the negotiations aimed at producing a Chemical Weapons Convention, with the consequence that this book is confined almost entirely to technical matters. My decision at least means that I write of what I know about; this is not the result of a professional writer's studying a topic and producing a more or less accurate account of it but, rather, an exposition of matters that have occupied my working life for some twenty-five years. The opinions and statements are, however, my own and not those of the Australian Department of Defence, with which I am affiliated. This book has no official endorsement as a statement of policy or intent of the Australian government. Nevertheless, I obviously owe much to my colleagues at work, principally Shirley Freeman (now retired) and Bob Mathews, both of whom have spent much time with diplomats achieving a harmony between the dictates of science and the aims of policy.

Last, I wish to inform, not to convert. Science can have a place in society only when it is understood by all that the determination of fact must be independent of the adoption of policy. How you interpret the information given here is your affair; my job is complete when the facts are given.

HUGH D. CRONE

Melbourne
April 1992

1

Introduction to a problem

The problem of chemical warfare has not been one of the major popular concerns of the late twentieth century, yet it has aspects that relate to, or impinge on, matters that have occupied public debate. Much more time has been spent discussing environmental problems, nuclear power generation and nuclear weaponry, chemical contamination of foodstuffs, applications of biotechnology and in vitro fertilization. Chemical warfare would result in environmental pollution; it is seen as a cheap substitute for nuclear war; toxicological questions relating to whether or not chemical warfare has occurred do have similarities to questions of contamination in foodstuffs; and biotechnology may make easier the production of one class of chemical weapons fill, the toxins. I have not yet thought out a relationship with in vitro fertilization, so score me four out of five. The real interest in the chemical warfare problem, however, is in whether an international convention can be negotiated that will ban it forever. This specific attempt at disarmament will need considerable technical input, and therefore I see it as the great test of science cooperating with diplomacy to solve a social problem. This is why I believe the matter should become prominent in public debate, for if we fail here, then the future for other disarmament measures, which will of necessity also be technically complex, is black. It is therefore more than a question of chemical war or chemical peace. It is a test case for the gradual limitation and strangling of war by successive proscriptions of the various means of waging it. But why is it chemical disarmament that has to carry the banner of disarmament? Because, as I hope you will read later, chemical weapons are the most difficult to control, presenting problems that exceed those with nuclear weapons. Public interest in this topic of the elimination of chemical arms has increased since the 1991 war in the Middle East and the investigation of the Iraqi stockpile by the UN Special

Commission, yet the technical and scientific aspects are little understood. Further, there is little awareness of the importance of this particular arms control attempt, which is seen as a side issue to the removal of nuclear weapons and the limitation of conventional forces. Perhaps it is minor in relation to military threat, but it is major in the prospects for arms control generally.

An arms control treaty or convention needs input from politicians, diplomats, jurists, military specialists and scientists. The politicians make the broad decisions, then the diplomats negotiate conditions acceptable to all sides, taking advice from the military on the possible scenarios and from the scientists on what is technically possible; next the diplomats commission the lawyers to draft the requisite clauses in legally meaningful terms. My only expertise is in science, so that what follows is my interpretation of the scientific aspects of chemical disarmament, in which I have taken care not to step on political, diplomatic, legal or military toes. For the occasional light stamp, I tender my apologies in advance.

The scientific and technical aspects of chemical disarmament have been well recorded in detail before, mostly in documents of limited appeal. Therefore I do not regard this book as a complete treatment of all aspects of the matter, indeed I could hardly claim this in such a short work. No, this book is intended as a key to that problem, a short and transparent introduction that will make clear the main issues and give the reader the sources of further information. Even then, I have kept references to a minimum to avoid making the book appear to be a formal text. There are references at the end of each chapter, with suggestions for further reading, and in addition at the end of this chapter I have listed two series of volumes on chemical warfare published by the Stockholm International Peace Research Institute (SIPRI), which has organised studies and conferences on the question of chemical warfare, in association with other bodies such as the Pugwash Conferences on Science and World Affairs. Their studies contain much useful information on many aspects of chemical disarmament.

This book is therefore an introduction to the scientific aspects of chemical disarmament intended to inform the layperson who may be interested in that topic specifically, or more generally in how science is bound into the affairs of our society. The reader may be interested in a more general introduction to disarmament questions, including the politics and diplomacy. This can be found in the work by Alva Myrdal,[1] who describes the disarmament scene in the 1970s, including the discussions on chemical disarmament. A full description of the negotiations towards a Chemical Weapons Convention, complete to the end of 1989, is given by Bernauer[2] for those readers interested in the detail both diplomatic and technical.

In the second part of this introduction I wish to discuss briefly some concepts that are necessary to understand the following chapters, beginning with the concept of toxicity. A chemical is toxic if it produces any adverse effect on a living organism that has been exposed to it. A toxic effect can thus be anything from a slight reduction in growth rate to an annoying irritation in the nose to quick death. Thus it is often necessary to specify a more definite endpoint such as lethality, incapacity or some other functional decrement. Any chemical whatsoever if given in sufficient dose will produce some kind of toxic effect, so that all chemicals are potentially toxic. So does the adjective "toxic" have any meaning? Yes, but only if a dose or some expression of concentration or quantity is also given. This is the principle of Paracelsus: "All things are poison, and nothing is not poisonous. The dose alone determines whether a thing is a poison." That is a free translation of Paracelsus's sixteenth-century German. Anyone setting out to choose a chemical for use as a weapon will obviously select one that has marked effects at low dose, the effect being usually, but not always, death. To put relative toxicity in a perspective, I have given some representative figures in Table 1.1. This shows chemical warfare agents of the nerve-agent class as the most toxic (effective at lowest dose) and then chemicals of lesser toxicity changing by factors of ten or more. Citations of further discussion on toxicity can be found in the References section of this chapter.[3]

A very toxic chemical, however, by this virtue alone is not always suitable for offensive use in war—a potentially very toxic chemical is of limited use if it cannot be spread around the battlefield and then enter the human body to kill or harm the victim. The physical form of the toxic chemical is important, and in general volatile liquids have been selected for war. This matter is further considered in Chapter 6, in relation to the natural protective mechanisms of the body.

It is commonly supposed that one group of chemicals can be clearly identified and labelled as chemical warfare agents. In fact, a wide variety of candidate warfare agents can be selected from an almost infinite pool of chemical possibilities. In 1918, for example, the British had succeeded in selecting 150 candidate chemicals, after three years of work. Today the field for selection is much broader, particularly with the advent of the organophosphorus and carbamate esters, and an increased understanding of how simple chemicals control the complex functions of the brain. There is no identifiable or cataloguable (if you will allow the term) list of chemical warfare agents, for only a functional definition can be given. The draft text of the proposed convention copes with this situation by a broad definition under Article II, which includes an element of intent, then attaches schedules

Table 1.1 *Acute toxicity estimates for a variety of chemicals*

Chemical	Dose[a]	Factor
Nerve agents Pufferfish toxin	10 μg/kg	1
Nicotine Hydrogen cyanide	1 mg/kg	100
2, 4–D herbicide	100 mg/kg	10000
DDT insecticide	300 mg/kg	30000
Malathion insecticide Aspirin	500 mg/kg	50000
Alcohol	4 g/kg	400000
Water	10 g/kg	1000000

[a]The dose cited is that causing death in a short time to adult persons. The route of administration is by mouth, except for nerve agents, pufferfish toxin and water, in which the route is by injection into a vein. The water dose is the author's estimate. All figures are approximate, as most are derived from accidents in which the exact dose was not known.

of specific chemicals and families of chemicals, which are either currently assessed as potential warfare agents, or as the precursor chemicals necessary to make the former.

The definition of chemical warfare agent has been used very broadly in some texts to cover almost any chemical used in, or associated with, war. Thus herbicides (prominent in the Vietnam War) and tear gas have sometimes been included in the definition; as a result that definition, in my opinion, is in danger of becoming quite indefinite. In this book I use the term "chemical warfare agent" to cover chemicals of high human toxicity in field doses, directed towards personnel and capable of causing permanent or semipermanent casualties—that is, removal of the victims from fighting duties for weeks or longer. Herbicides and tear gases are carefully screened to ensure that the long-term human toxicity is minimal, the reverse process compared with selection of a warfare agent. We cannot put two opposing functional groups of chemicals into one category, but we may usefully put the non—personnel-directed chemicals into other functional categories, for example, use of herbicides in certain circumstances may fall within the proscription of the United Nations 1977 Convention on Environmental Modification Techniques. This leaves me with the problem of classifying tear gases, which I know most other commentators see as clearly being warfare agents when used in war. Therefore my personal difficulty just noted (long-term minimal toxicity of a tear gas) can be resolved by permitting the use of tear gases (riot-control agents) in domestic law enforcement but not in warfare. Such a proposal is under discussion for inclusion in the draft text. I discuss this matter further in Chapter 5 and in my concluding remarks in Chapter 11.

I end this introduction with a quite different observation. The science of peace and the science of war share much common ground, for the facts are the same whether the application is to attack or defend. However, the applications are clearly distinguishable; defense is not attack, despite a constant muddying of the distinction between defensive applications and offensive developments. If we are to have a Chemical Weapons Convention, it will largely be founded on the efforts of defence scientists in a large number of nations, east and west, north and south, but the final result will come from political decisions influenced by perceptions of what you, the people, want. I believe that we scientists can deliver, so the result is in your hands.

References

1. Myrdal, A. (1977). *The Game of Disarmament.* Manchester: Manchester University Press.
2. Bernauer, T. (1990). *The Projected Chemical Weapons Convention: A Guide to the Negotiations in the Conference on Disarmament.* New York: United Nations.
3. Crone, H. D. (1986). *Chemicals and Society: A Guide to the New Chemical Age.* Cambridge: Cambridge University Press.

A series of six volumes produced by SIPRI is entitled *The Problem of Chemical and Biological Warfare* (published by Almqvist and Wiksell, Stockholm, on dates between 1971 and 1975): I. *The Rise of CB Weapons.* II. *CB Weapons Today.* III. *CBW and the Law of War.* IV. *CBW Disarmament Negotiations, 1920–1970.* V. *The Prevention of CBW.* VI. *Technical Aspects of Early Warning and Verification.*

Volumes in the series of SIPRI Chemical and Biological Warfare Studies (published by Oxford University Press, on dates between 1985 and 1990) include:

No. 1: *Effects of Chemical Warfare: A Selective Review and Bibliography of British State Papers,* by Andy Thomas.

No. 2: *Chemical Warfare Arms Control: A Framework for Considering Policy Alternatives,* by Julian Perry Robinson.

No. 3: *The Detoxification and Natural Degradation of Chemical Warfare Agents,* by Ralf Trapp.

No. 4: *The Chemical Industry and the Projected Chemical Weapons Convention.* Proceedings of a SIPRI/Pugwash Conference, volume I.

No. 5: *The Chemical Industry and the Projected Chemical Weapons Convention.* Proceedings of a SIPRI/Pugwash Conference, volume II.

No. 6: *Chemical and Biological Warfare Developments: 1985,* by Julian Perry Robinson.

No. 7: *Chemical Weapon Free Zones?*, edited by Ralf Trapp.

No. 8: *International Organization for Chemical Disarmament*, by Nicholas A. Sims.

No. 9: *Non-Production by Industry of Chemical-Warfare Agents: Technical Verification under a Chemical Weapons Convention*, edited by S. J. Lundin.

No. 10: *Strengthening the Biological Weapons Convention by Confidence-Building Measures*, edited by Erhard Geissler.

No. 11: *National Implementation of the Future Chemical Weapons Convention*, edited by Thomas Stock and Ronald Sutherland.

A recent study of the broader issues of the present topic is: V. Adams (1990). *Chemical Warfare, Chemical Disarmament.* Bloomington: Indiana University Press.

2

The science of peace and of war

War has been a part of our history, and that of many other cultures, from the beginning of recorded time. Whatever the reason for this, the result has been that the preparation for war, or the defence against it, has become part of our social structure. Anglo-Saxon society had a class of male warriors as a basic part of its culture, as evidenced by the earliest literary works. Thus the Old English poem "The Battle of Maldon" (A.D. about 1000) describes the practical expression of the warrior code: The thanes lay dead around the slain leader Byrhtnoth, as this was the required payment for their social status and their places in the mead hall in times of peace. This culture of the warrior received even more formal expression in medieval times, when the knight owed a fee of military service to his lord, and the latter likewise to his king. You may believe that this concept disappeared with the dissolution of medieval society, but I believe it persisted until at least the cataclysm of 1914–18. One cannot gain any understanding of that episode of senseless and repeated slaughter unless one considers there was a strongly rooted feeling of military obligation. A generation of young men was sent to Flanders to serve as junior officers and lasted—how long? Three months, six months perhaps, they survived until the machine guns cut them in half as they led another charge against barbed wire. My father was born in 1899, a fortunately late date for him, as he was a young lieutenant on the way to war when the armistice was signed in 1918. He would say that the young men did not see their service as futile but wished to do their bit for their country. In the United States, the concept of a warrior caste was modified by frontier conditions to that of a self-reliant citizenry who would look after themselves, a militia that did not have the more rigid social position of the British officer class.

ONE OF THE "MAXIMS" OF CIVILIZATION!

OLD AND NEW

"THINK of the glorious Mottoes," said a Major of the
old school. "'*Nil Desperandum.*' 'Death or Victory,
'England Expects,' and so forth!" Replied his friend,
the modern Captain, "Bother your Mottoes! Give us the
'Maxims'!"

Figure 2.1. This cartoon from Punch of 2 December 1893 shows how technological
superiority in war was appreciated in the colonial era.

Many other cultures, but not all, have had this social organisation for war
deeply embedded for millennia: the Pathans and other tribes of the north-
west frontier of India, the Plains Indians of North America and the Fuzzy-
Wuzzies of the Sudan, Tommy Atkins' most admired foe. The Japanese had
a feudal society until recent times, which helps to explain their barbaric
conduct in the Sino-Japanese and Pacific wars of the 1930s and 1940s. Like
it or not, we have a society in which the expectation of war has supported
a military caste; to some this is a necessity for survival, to others a provo-
cation for further conflict.

Against this background, it is not surprising that science and technology have been pressed into military service. In fact, successful conquest has been largely dependent on superior technology (Fig. 2.1), as many instances over the ages testify. There are legends in which iron had the power to scare off the fairies of the moorlands, who can be regarded as the users of bronze who were defeated and displaced by the wielders of the superior iron swords and spears. The extraordinary exploits of Cortes and the Pizarros in Mexico and Peru must have largely been due to the possession of firearms, although the effects of the latter may have been more on the psychology of the natives than on the actual numbers of people killed. The outcomes of the colonial wars in Africa and the Indian wars in North America were determined by Martini-Henry, Remington and Colt. The technically superior combatant was beaten only when a gross military error was made, as when the British columns were caught on the march at Isandlwana by Zulu impis, who promptly cut them up. Perhaps the most graphic demonstration of the inevitability associated with technical superiority was in the naval battle of the Falkland Islands in 1914. When the German raiding squadron of the Graf von Spee neared the islands to refuel, the Graf could see the superstructure of British Dreadnought battlecruisers above the horizon. At that moment he knew he was lost, as the guns of the Dreadnoughts had greater range than his own. The German ships scattered, but most were lost as the British pursued.

The war of 1939–45 provided many instances of the use of science and technology in war, and has generated a number of debates on the actual value of the great variety of devices invented.[1,2] Of course, the atomic bomb was the most controversial product of that conflict.

Among technical contributions to war, it is not surprising that chemistry was pressed into service. Chemistry of course enters into many applications—explosives, metals, other fabrics, medicines—but of most interest to us is the use of chemicals that are directly toxic to humans. The history of this application of chemistry is the theme of the next chapter.

The science of peace has, to understate the matter, had a rough passage through history. Very few cultures have been entirely pacific in outlook; the Australian Aborigines and the Indians of the Pacific coast of North America were so, at least when the latter could keep away from the scalp hunters of the plains. Other cultures have had religious beliefs of pacifist nature, such as Buddhism and the various cults in India, but often these beliefs seem to have overlaid essentially turbulent cultures. The sectarian violence in India is an embarrassing reality to cultures professing nonviolence. The Christian tradition has been similarly ambiguous with respect to war; most of the wars have had religious blessing—in fact, many were caused by religious schism. It is not surprising, therefore, that the science of peace, in contra-distinction to that of war, has not had much encouragement. It is only in recent times that the study of human aggression and the other causes of war has been conducted in a scientific manner. I do not wish to discount the vast amount of wisdom accumulated over the years by those who have been dedicated to peace and who have struggled to record and analyze the forces that drive humanity to inhumane acts. By a scientific manner, I mean an objective study of human behaviour in which most of the variables are controlled, an extremely difficult undertaking given the complexity of the factors that influence behaviour.

The modern approach to the problems of aggression and of war origi-nated with the studies of animal behaviour by researchers including Tin-bergen, Lorenz and Hinde.[3] This topic has become more defined in scope to cover the physiological, ecological and genetic aspects of behaviour, and has been named ethology. The attempts to explain the origins of war have had to draw on human psychology, comparative psychology and the biol-ogy of human societies together with ethology in one of the most complex studies that our society must undertake and master. As science solves the more pressing problems of the world (disease, hunger, shelter) we are left facing the most difficult, which is the control of ourselves rather than of Nature. In theory at least we can defeat the horseman who is Famine, delay the one who is Death or Disease, welcome the one who is Christ, but the fourth horseman of the Apocalypse is War and yet unbeaten.

Ethology has generated a major question, still unresolved, which has split its students into two camps. Is aggressive behaviour genetically deter-mined, or is it acquired in early life from the social milieu in which the in-dividual is raised? If the first proposition is true, then there appears to be an inevitability in the occurrence of aggression. If the second alternative is correct, then society may be able to reform itself by paying careful attention to the upbringing of its children. The inbuilt or genetic origin of aggression is argued by Lorenz,[4] whereas the case for acquisition is defended by Hinde and others.[5] I have included the Seville Statement on Violence as an appen-dix to this book, as it summarises the beliefs of the latter group.

If you reflect on these two opposing views, you will see some of the dif-ficulties in deciding between them. Aggression and war are so rooted in many societies that one is led to believe that they must have some common, genetic origin in mankind's nature. But then one has only to attend a school football match to see how parents will urge their six-year-old sons on to acts of aggression and reward success at competition. Much is obviously learnt from one's peer group in society, whatever that group may be. But is it all learnt? Our pet duck will confront any visitor to our backyard, and the cats are careful to avoid her. This behaviour is presumably beneficial for the survival of ducks, and must be inherited, but does it have relevance to aggression within species? As a postscript, I add that a fox took our duck one night, which shows that formal aggressive behaviour only works if both parties know the rules. The fox obviously had a different set of rules.

The argument of Lorenz and his supporters is not so much that aggression is a simple inherited trait in humans as that it is part of an inbuilt complex of reactions to specific situations that formerly occurred and contained countering mechanisms to the aggressive acts. Modern society has removed many of these mechanisms that inhibited the continuance of aggression, so that what were once behavioural traits useful to the species are now threatening our continued existence on this planet. The opposing view is that war is a peculiarly human activity, with no precedent in animal behaviour, and therefore a learnt, cultural response rather than an innate, biological one.

The connection between aggression by an individual and a society's going to war must also be a complex one, which will require much study. It is not hard to see why the scientific study of war (or of peace) is unlikely to yield results in the short term. Physical scientists may not realise just how difficult the problem is, for they have made extraordinary advances. We can put satellites into precise orbits, make electron microscopes that can see large, single molecules, and make small chips that can process masses of information. This, however, is easy science. It is the science of behaviour that is really complex.

The purpose of this section is to show that the devices of war can be easily made; the missiles, explosives, toxic chemicals are well within our technical grasp. Peace is not; if we are to achieve chemical disarmament, or the banning of any other type of weaponry, then we must work for it piece by piece, inch by inch. To do this we must first understand the problem in technical terms, which is the purpose of this book.

Negotiation has achieved some limitations on war in the past, notably the conventions relating to the treatment of prisoners of war (1949), environmental modification (1977) and excessively injurious or indiscriminate weapons (1981). It is true that such agreements are fragile—for example, Iraq had no hesitation in breaking the Geneva Protocol—yet together they may limit the horrors of war.

In the following chapter we will look more specifically at the use of chemicals in war, starting with a short history.

References

1. Jones, R. V. (1978). *Most Secret War.* London: Hamish Hamilton.
2. Macksey, K. (1986). *Technology in War.* New York: Prentice-Hall.

3. Hinde, R. A. (1966). *Animal Behaviour.* New York: McGraw-Hill.
4. Lorenz, K. (1966). *On Aggression.* New York: Academic Press.
5. Groebel, J., and R. A. Hinde (eds.). (1989). *Aggression and War.* Cambridge: Cambridge University Press.

3

Chemistry and chemical warfare

This brief review of the history of chemical warfare forms a background against which to set the current problem of chemical disarmament. Any advance in technology has been followed almost immediately by the application of that technology to war. This is as true for Chemistry as for any other science and its related technology, so that this chapter is in effect a history of chemical technology, with the emphasis shifted towards war.

As far as the West is concerned, the recorded history of chemistry begins in the early years of the Islamic era, before A.D. 900 (A.H. 280), although there is a separate recorded history in China and perhaps elsewhere. Here I ignore separate developments in chemical technology that may have occurred in more isolated cultures. At that time alchemy and chemical technology were not distinguished, both being embraced in the Arabic term *al-Kimya*. As with many other scientific and cultural advances arising from Islam, it is not easy to determine how many were Arabic innovations and how many were translated from other cultures (e.g., China or India). This is because Islam formed an enormously extensive cultural entity stretching from West Africa to Indonesia–a great diffusion mechanism for knowledge. This diffusion is exemplified by the wanderings of Ibu Battuta from his home in Fez (Morocco) to West Africa, then Egypt, Arabia, India and probably China, between A.D. 1325 and 1353. It has been speculated[1] that the origin of the Arabic *al-Kimya* is Chinese, from a word for gold. There is no doubt, however, in the chemical attainments of Islam, the main accomplishments being in smelting and metal refining, distillation (Fig. 3.1), the use of mineral acids and alkalis, soap and glass manufacture, pigments, dyes and other products. Of interest at the present time is the fact that petroleum from Baku and Iraq was distilled to provide fuel and for its solvent and medicinal properties.

13

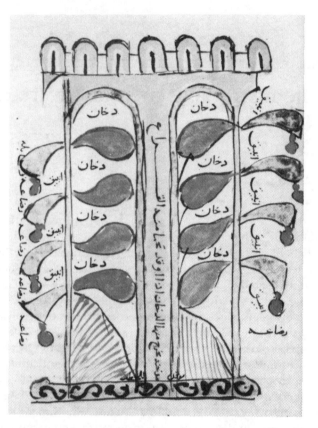

Figure 3.1. Distillation of perfume from flowers by hot air. The air from the fire (bottom) rises to heat the flasks within the furnace, and the vapours pass to the funnels outside, from which the condensed liquid slowly drips. Manuscript of al Dimashqi, fourteenth century. Bibliothèque Nationale, Paris.

This chemical technology was put to military use in the form of incendiary weapons and, later, explosives and gun powder. Smoke would be an inevitable accompaniment of the fire weapons, but there is no evidence of the deliberate generation of asphyxiating gases. The incendiary substances were petroleum products, resins, sulphur and saltpetre, often used in fire pots or pottery grenades hurled by catapults, or carried in rockets or torpedoes. Later on, the development of cannon (first reported in the thirteenth century in North Africa and Spain) became of supreme military importance, allowing the Ottoman Turks finally to blast their way into Constantinople in 1453. One gun is reported to have had a bore of 820 mm; today the normal heavy artillery bore is 155 mm. In the narrow sense of the use of gases and liquids strictly for their toxic effect, the Islamic chemists did not develop chemical warfare, but they certainly created the technical basis for it (and for a multitude of powerful, constructive uses of chemistry).[2]

The limits of the Turkish Empire were defined by the failure of Suleiman to take Vienna in 1529 and Malta in 1565, but this circumscription of temporal power did not limit the spread of knowledge, which steadily diffused to the West through trade contacts (Venice and Genoa), through refugees from the crumbling Byzantine Empire, and through the mixture of Christian and Moorish influence in the Iberian Peninsula. We are traditionally presented with the image of the alchemist as the originator of chemical knowledge in the West, but I believe this is a romantic distortion of reality. The quest to find the means to turn base metals to gold, or Dr. Faustus's conjurings of the devil, make good tales (and let's recognise Marlowe's precedence over Goethe), but many practical men were interested in the uses of the base metals themselves. Chemical technology thus has roots in metal refining, herbal medicine, and the production of household requirements such as soap, glass, paper and tanned leather. Much of the primary industry of Western Europe from the thirteenth to the fifteenth century was in the hands of the great monastic houses, which for example exported wool from the north and east of England to other houses in Flanders and France for the production of cloth. Industry gradually became secularised as towns developed and became more independent. Initially, however, there seems to have been little development of chemical technology, so that although artillery rapidly appeared in the West (e.g., at Crécy in 1346), it was as a received technology, not exploited locally. There was even less Western development or use of flame and incendiary weapons. Thus for several centuries Western military technology concentrated on the development of guns, but not on anything leading to poisonous gases or liquids for warfare.

The modern Western development of chemical warfare occurred within the past century with the growth of a large chemical industry, whose own origins are traceable to the roots we have discussed. In Chapter 8, I give a series of snapshots of the recent development of the chemical industry to illustrate the way it has pervaded modern life. Here, I will discuss more specifically its development in relation to that of chemical warfare. By the late nineteenth century various chemicals were produced in quantity, substances such as chlorine, sodium hydroxide (caustic soda), ammonia and the mineral acids (sulphuric, hydrochloric, nitric, phosphoric, etc.). Also the synthesis of much more complex organic compounds—that is, compounds principally composed of a chain of carbon atoms—was actively being investigated after the successful development of the synthetic dyestuffs industry. This latter expertise was directed towards producing better military explosives, as the nitration of natural substances (e.g., guncotton) was followed by the synthesis of nitro compounds (nitroglycerine, nitrotoluene), but not towards poisonous gases for war. The nineteenth-century suggestions for use of asphyxiating gases in war were haphazard ideas, not related to the growth of industry.[3]

However, chemical industry had provided the means for chemical war, but this alarming potential lay latent and largely unrealised until one man, Fritz Haber, organised a German effort that led to the use of poison gas in 1915. That some thought about the use of toxic chemicals was present in the minds of government and the military is shown by the prohibition on the use of projectiles whose sole object is to diffuse asphyxiating gases in the Hague Convention of 1899, and by tentative experiments with shells and grenades containing tear gas.

The beginning of chemical warfare in the modern era, that is, the use of toxic chemicals on a wide scale, can be recorded precisely. It was at 17.30 hours on 22 April 1915, near Ypres, in Flanders, probably the most expensive piece of real estate in terms of human lives in the world. At the time mentioned, the Germans released 150 tons of chlorine from gas cylinders against the Allied lines. The gas attack was a surprise and effective in its own terms, but not properly followed up by the German military, and was thus not an overall military success. Haber himself is reported as saying that chlorine had been used earlier that year on the Russian front, with even less success, as the chlorine combined with snow to form chlorine hydrate, a nonvolatile complex. Thus the initial use of chlorine, and later phosgene, released as a cloud had military and psychological effects on the Allies, but the Germans also suffered gas casualties due either to encountering pockets of gas when advancing against retreating Allied troops or to wind changes after release of the gas. By 6 August 1915 the Germans had released 1,200 tonnes of gas, over two-thirds of this against the Russians in the east. The use of gas was not markedly successful, and one may have thought that it would be abandoned then, but the genie could not be withdrawn into the gas bottle; initial use prompted crude Allied defensive measures and retaliation in kind. This in turn prompted further German research into more effective gases and delivery means, and the cycles went round and round until 1918 and the armistice.

The circumstances that led to the use of chlorine in 1915, and the technical aspects of the use of chemical warfare in World War I are well described in a book by L. F. Haber,[3] which also discusses the key role of his

Table 3.1 *The major toxic chemicals used in World War I*

Chemical	Date of first use	Relative toxicity [a] (chlorine = 1)	Density of vapour relative to air
Chlorine	April 1915	1	2.45
Phosgene	1915–16[b]	20	3.4
Hydrogen cyanide	1916	12	0.9
Diphosgene	1916	–	6.9
Chloropicrin	1916	–	5.7
Mustard	July 1917	40	5.5
Lewisite	Not used	40	7.2

[a]The relative toxicities are based on the inhalation lethality of the gas or vapour.
[b]Phosgene mixed with chlorine was used in 1915, but wide-scale use commenced in 1916.

father, Fritz Haber. That chlorine was the chosen gas was inevitable from the need and from the state of the industry. The need for chemical warfare is for a chemical in good supply, because the chemical weapon is an area attack weapon rather than a targetted one, and is consequently wasteful of the resource. The toxic chemical that industry could best supply was chlorine, a key chemical in the chlor-alkali industry, which is still at the heart of all chemical industry.

The reader should consult Haber[3] for details of the chemical war from 1915 to 1918, and an account of the British reaction from the military aspect is given by Foulkes.[4] A good summary is provided by Hartcup.[5] Here I will only provide a sketch, extracting what I consider to be the essential elements.

First, the development of weaponry proceeded rapidly after April 1915. The term "chemical weapon" includes both the toxic chemical and the means by which it is delivered on an enemy (see Chapter 5). Only the principal toxic chemicals used are given in Table 3.1; many more were used in a minor way or were investigated for military use but not employed. The table shows that within one year chlorine had been supplanted by more toxic chemicals. You may wonder why I have included the density of the gas or vapour relative to air in the table, but it is fairly obvious that a gas lighter than air will disperse more rapidly than a denser one. Chlorine demonstrated this well at Ypres. In the first attack the upper edge of the chlorine cloud rose to 10–30 m, but followed the ground contours and remained in pockets of low ground. On 24 April, a second chlorine cloud, released at two o'clock in the morning, travelled as a compact zone of 2 m in height (presumably because the ground was cooler than on the afternoon of 22

April). Hydrogen cyanide, which is lighter than air, never proved of any use in war, although the French employed it for a time. Substances not readily volatile—those that remain liquid for long periods—are not much use for cloud gas attacks (e.g., diphosgene, chloropicrin). A combination of toxicity with volatility and vapour density is found in phosgene, an observation that the British were quick to make. Phosgene became the favoured British gas, and accounted directly or indirectly for the majority of the gas fatalities, at least on the western front. Phosgene is peculiarly nasty and insidious in its effect on the person who breathes the vapour. There may not be any effect for an hour or so if the dose is moderate, then the person has increasing difficulty in breathing as the lung tissue is slowly destroyed and fills up with body fluid. Death, which is slow in coming, is by asphyxiation. The patient may lie near death for days, in which case the damaged lungs are ideal harbours for bacterial infection. The high mortality associated with phosgene poisoning was partly due to pneumonia setting in at this stage, a disease for which no antibiotics then existed. The British soon discovered the nature of phosgene poisoning; there is a story of one soldier who was engaged in setting up and checking phosgene cylinders before an attack on the Germans. He felt all right after the job, but died a day later; one or more of the cylinders had been leaking.

A second phase of development occurred with the introduction by the Germans of mustard "gas" in 1917. Call it what you will, sulphur mustard, Hun Stuff, Hot Stuff, H.S., Lost, Yellow Cross or whatever, mustard is not a gas but a rather involatile liquid. "Gas" and "gas warfare," however, were firmly fixed in the common vocabulary and remain there to the present time. Sulphur mustard slowly destroys any living tissue it contacts. If the liquid contacts skin, it destroys the skin. If the vapour is inhaled, it destroys the lining of the lung. The liquid has sufficient vapour pressure to make the vapour a threat, although it is as mentioned much less volatile than water, kerosene or the like. If mustard kills, it kills slowly, but the main effect from a military point of view is to create many casualties with horrific injuries. Mustard from its first use was quickly perceived by all sides to be the most effective war chemical.

Organic compounds of arsenic were thoroughly investigated from 1915 to 1918, because many had unpleasant properties. One compound, DM or Adamsite, developed by the United States towards the end of the war, causes violent and uncontrollable vomiting. It was used later as a riot-control agent but was withdrawn from that role long ago because of its low safety margin (effective dose too near the lethal dose). The Americans also developed Lewisite, an arsenical compound with properties like mustard. Production of Lewisite was being initiated in the United States when the

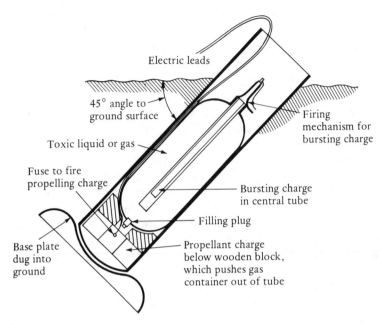

Electric leads

45° angle to
ground surface

Firing
mechanism for
bursting charge

Toxic liquid or gas

Fuse to fire
propelling charge

Bursting charge
in central tube

Filling plug

Base plate
dug into
ground

Propellant charge
below wooden block,
which pushes gas
container out of tube

Figure 3.2. Diagrammatic representation of a Livens projector. The cylinder of gas is shown in the firing position, within the firing tube dug into the ground.

armistice was signed. It has since been frequently put into munitions as a 1:1 mix with mustard; by itself it is not regarded as a very effective chemical warfare agent and the Americans are not now enthusiastic about its value, but some nations have stockpiles of it.

The development of the technology of chemical poisons was followed closely by that of the means of delivering the chemical on target (accepting that precise hits were unnecessary). Arguments about the best methods continued to 1918; the disadvantages of cloud gas (massive release from cylinders in your own lines) were immediately appreciated as the wind changed to blow the gas back onto friendly troops. The alternative was to package the chemical and send the pack over the enemy lines to burst and release the chemical there. The artillery shell is the obvious package but has the disadvantage that the chemical payload is low (see Chapter 5). Mortars were adapted to fire chemical bombs, and the ingenious Major Livens invented the projector named after him, which when fired en masse created an effective cloud of phosgene at distances to 1.5 km. The device (Fig. 3.2) had the advantage of being relatively simple and easy to conceal before use, as it was buried in the ground. Packaged chemicals did allow another dimension in the distribution of chemicals, as I shall explain. The criterion for selection of a cloud gas is that it is a gas at normal atmospheric condition. Being a gas, it is dispersed by wind, and therefore does not persist over enemy lines, except to the extent that denser gases remain for a short time in hollows and trenches. A liquid, however, if carried to the target in a container, then ejected to contaminate an area of ground, will continue to release vapour into the air until the liquid is exhausted. The effect is much more persistent than with cloud gas. Further if the liquid has a toxic effect through the skin, then there is a contact hazard to people as well. Whereas one could continue to argue that cloud gas attacks with phosgene were useful, as Foulkes did, one could not do so for mustard. This persistent liquid required a delivery package to get it to the enemy, and the Germans used artillery shells. "Gas" warfare ceased to be gas warfare around 1917, but we are unable to get the term out of the vernacular. However, you should remember that chemical warfare is "liquid/vapour" warfare; solids have been used but are not readily absorbed by the body, except as aerosols (see Fig. 6.3).

Parallel with the developments in offence were frantic activities to devise and produce defensive measures. The immediate need was to protect the lungs, and after initial bad tries, some effective respirators or gas masks were produced. An early Australian example was designed at the University of Melbourne (Fig. 3.3). This was "guaranteed to be efficient in air containing chlorine, bromine, sulphur dioxide, or other poisonous gases of acidic character. Total cost (Melbourne prices) 7 shillings." The British also found it to be effective against phosgene and hydrogen cyanide, but declined to make further orders because their product under development would meet all requirements. This latter is a doubtful statement because at the time (October 1915) the United Kingdom was playing with the design of what became the "PH Helmet"—not a startling success, but the Poms (the English) could not admit the superiority of a colonial product. Gas masks were produced in quantity in poor conditions (Fig. 3.4), but quality-control procedures were introduced and by 1918 the gas mask was a reasonably effective and reliable means of defence, at least for the British and Germans. The use of persistent liquids such as mustard necessitated protection for the skin as well as for the lungs, and various countries experimented with protective clothing. They immediately encountered the problem that continues to this day, namely that a high degree of protection equates to a great encumbrance on movement and comfort. Gas capes of rubber or oiled fabric were devised, but by 1918 there was no accepted general issue of anti-gas clothing.

When the armistice was signed in November 1918, what was the outcome of this frantic creation of chemical warfare? In military terms, perhaps nothing: The stalemate on the western front was not resolved by chemicals. In terms of military technology, an enormous field had been opened up. The effort expended on chemical warfare research was extraordinary; it is estimated that in the United States, which entered the war in April 1917 with no experience of this kind at all, one in ten of the chemists was working directly on chemical warfare research by 1918, mainly as unpaid volunteers in universities.[6] The research stations created in the United Kingdom (Porton Down, 1916), the United States (Edgewood Arsenal, late 1917) and other countries consolidated the hurriedly gained wartime knowledge, and the many university researchers pressed into wartime service reverted back to their peaceful studies in synthetic chemistry, toxicology, respiration physiology, or whatever. A vast number of chemicals had been examined for their utility in war, and a few retained in munition: mustard, phosgene, Lewisite, chloracetophenone (a tear gas) in the United States, mustard, phosgene and bromobenzyl cyanide (another tear gas) in the United Kingdom. The two tear gases were nearly useless in war, and Lewisite decomposed so rapidly in humid air as to be of little value. Arsenical compounds diffused as "smokes" were still of interest, and the Japanese in World War II, for example, had a range of smoke candles and similar devices. Thus phosgene and mustard were the effective weapons to be passed from one world war to the next, gifts not enthusiastically received by most sections of the military who did not understand chemicals and saw them as unchivalrous obstacles to the ancient art of blowing enemies to bits.

In terms of arms control, there was also an outcome, but of a partial and unresolved nature. Obviously, the prominence of chemical warfare ensured that the topic received much attention when the nations tried to draw up treaties to stabilise international affairs. The Treaty of Versailles (1919) that formally ended the war with Germany forbade the use, manufacture and importation of poisonous gases in Germany. The abortive Washington Naval Treaty of 1922 had attempted to introduce a ban on chemical weapons, but the treaty did not enter into force. Nevertheless the topic had been discussed internationally and there was a general will to ban chemical weapons. As a kind of stop-gap measure, the parties to an International Conference on the Control of the International Trade in Arms, Munitions and Implements of War, meeting in Geneva in 1925, agreed to a short expression of renunciation in the "Protocol for the Prohibition of the Use in War of Asphyxiating, Poisonous or Other Gases, and of Bacteriological Methods of Warfare." This Geneva Protocol is still the definitive statement on chemical weapons, but its limitations must be understood.

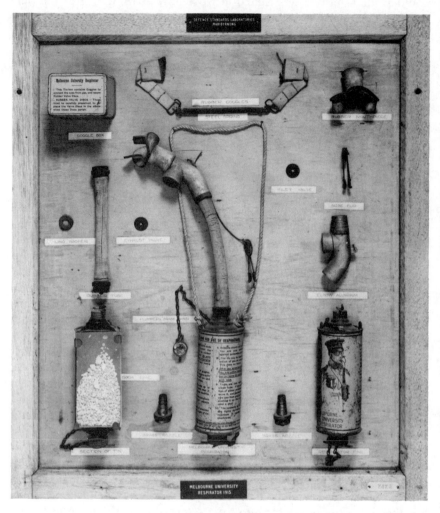

Figure 3.3. The respirator designed by Professors Laby and Masson of Melbourne University in 1915. Left: An original item on a display board. Right: Original diagram of the canister and mouthpiece, part of the manufacturing specification. From the records of the Materials Research Laboratory, Melbourne.

First, it is a formal restatement and extension of existing customary law and other prohibitions, such as the Hague Declaration 2 of 1899, so that it was not entirely novel, but more a means of correcting the breaches that had occurred from 1915 to 1918 by including all chemical weapons, not

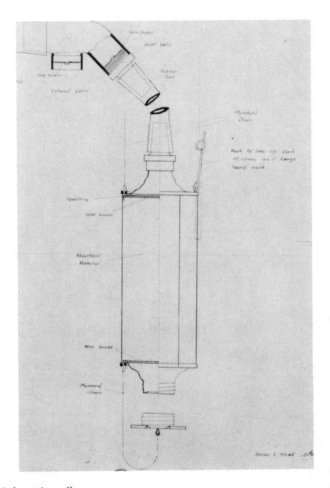

Figure 3.3 (*continued*).

just projectiles. It also for the first time specifically mentioned bacteriological weapons.

Second, it was not a well thought out convention with methods of implementation and enforcement written into it. The strong impression is that the nations expected such a convention to follow later, but at that time the United States moved towards isolationism and international resolve became weak and directions were muddied (the United States finally ratified the convention in 1975, a fiftieth birthday present).

Figure 3.4. These ladies are repolishing the eyepieces of early gas masks. Probable location is the Crowndale Works, Charrington Street, St. Pancras, London, in 1916. Students of industrial safety will note that, although working comfort is low, the machine belts are guarded, the polishing wheels have partial shields and the workers have coveralls and wear their hair well tucked up. From an original print held at the Materials Research Laboratory, Melbourne.

Third, approximately a quarter of the nations that ratified the protocol or acceded to it made the reservation that they would respect it only as regards countries who were also signatories, and that they would cease to be bound by it if they were first attacked with chemicals or bacteria. Reprisal in kind after attack would have been acceptable anyway under customary international law, but this reservation added to the view of the protocol as being only a prohibition against first use of chemical warfare. It does not prohibit a country from acquiring chemical or bacterial weapons as a standby against attack. Ireland withdrew its reservation to the Geneva Protocol (in 1972) and declared its resolve not to use chemical weapons even if attacked with them, and Australia and other nations followed this lead.

The protocol is thus weak, but it is sufficient to allow the denunciation of those countries that have used chemical weapons: Italy in Abyssinia (protocol ratified 1928, broken 1935), Japan in China (signed 1925, broken 1938 or before but not ratified until 1970), Iraq in Iran (acceded 1931, broken 1983), perhaps others. However, after having being denounced, the offending nations have continued about their business quite cheerfully, leaving their international censors to weep over the impotent clauses of the protocol.

From 1925 we sweep forward from an uneasy peace to the threatening developments of Italian and German fascism and Japanese imperialism. In the late 1930s war was seen as inevitable, and the military were prepared for chemical war. There was also, however, the novel possibility of chemical attack on civil populations.

Civil defence authorities in the 1930s had two unknown factors to ponder over. One was the effectiveness of aerial bombardment by contemporary aircraft in general, and the second was the effectiveness of chemical attack on a civilian population.

The experience of air raids in World War I, although limited, had alarmed the civil authorities. England is said to have suffered 1,414 killed and 3,416 wounded,[7] and Germany 2,500 killed. It was not known exactly how destructive mass raids by more modern aircraft would be, but the attack on Guernica in the Spanish Civil War had heightened popular concern, so that in the late 1930s the defence of the civil population became a matter of great concern in the United Kingdom, Germany and other European countries. The English were to find out fairly soon how effective an aerial blitzkrieg could be, as Coventry disappeared under incendiaries in 1940.

There was no experience at all concerning the bombing of cities with toxic chemicals, so that speculation was even wilder on this matter. Nevertheless, the issues were discussed, and contemporary books on this topic[7,8] contain sensible analyses of the situation and descriptions of

Figure 3.5. The cover of this 1940 civil defence magazine illustrates the U.K. General Civilian Respirator, similar to the one worn by the author as a child. From the collection of the Materials Research Laboratory, Melbourne.

defensive measures. In the United Kingdom, the government organised the production of a gas mask for every person; some thirty million of the general civilian respirators (Fig. 3.5) were available by 1938. This was the model that I had as a child, in the early 1940s, in the cardboard box that I carried to school for gas mask drill, but I was not "smiling through." Other civil defence stores included eye shields or goggles, antigas ointment and protective clothing of oilskins, the latter only for ARP (air raid precautions) workers, firemen and others in an official capacity. I have in front of me a copy of the booklet "A.R.P. Handbook No. 1 (1st Edition) Personal Protection Against Gas, 1937," which seems to me to be a good, clear account of protective measures as then available, indicating that a great deal of thought had been put into civil defence by 1937.

One may wonder about the effectiveness of this equipment, and the basic instruction and drills that were given. Haber[3] speaks disparagingly of the civilian respirator, because it was ineffective against arsine and other arsenicals. This is true, but it was effective against the main threat, mustard and phosgene, and the arsine problem was corrected later by addition of an extra filter (a disc was attached to the outside of my filter by a rubber band, about 1942). One can also speculate about why toxic chemicals were not used by the Germans (or the British), but one contributing reason was that the opposite side was seen as having effective protection against chemicals; high explosives and incendiaries worked well. Another reason was of course the fear of retaliation in kind, as each side was well provided with chemical munitions. What many people do not appreciate now is the feeling in the 1930s that gas would inevitably be used in the coming war, and the enormous degree of preparedness, both offensive and defensive. It is almost miraculous that gas was not used. The military of course were as well prepared as the civilians, so that gas masks and gas capes were items of common issue.

So what of the military? All nations had chemical weapons, mainly containing mustard, phosgene or a mix of mustard and Lewisite (popular in Soviet Russia and Japan). All had gas protection, detection and decontamination systems, often poor at the beginning of the war, but rapidly improved after 1939. The effects of the principal warfare agent, mustard, on volunteer soldiers under different conditions was tested in Britain at Porton Down, in Canada at Suffield and in Australia under tropical conditions in Queensland. German, Japanese and Russian trials were also conducted, so that there is a great deal of information on the toxic effects of mustard derived from a war in which it was not used in anger. There were a great many false alarms about the use of gas, and a great deal of enemy equipment was captured and its efficiency assessed. Some of the alarms were comic, as when a mustard gas raid on an Australian position in New Guinea was suspected. A group of Australian soldiers were peeling parsnips in the open when a Japanese airplane flew overhead. A little time later, the Australians noticed that their hands were reddened and blistered. This was not, however, due to a shower of mustard gas, but to a component of the juice of parsnips known as psoralen, which sensitises skin to sunlight and can cause dermatitis.

The big development in offence was that of the nerve agents or organophosphorus esters by the Germans, and later by the British (see my account).[9] Unlike mustard and phosgene, the nerve agents were not deployed to the war fronts in munitions. One suspects that Hitler was not in favour of chemical weapons, and that he had greater hopes from the V–1 and V–2 rockets under development later in the war. In terms of delivery methods, the munitions were not sophisticated: aircraft bombs, aircraft sprays, artillery shells, mortars, projectors and so on.

Gas was therefore a nonevent in World War II, and at the end the nations destroyed much of the chemical weaponry, threw away the protective equipment and reduced their research effort on chemical warfare. All of the remaining effort was devoted to the study of the nerve agents, resulting in the synthesis of new ones (the V series of very toxic, nonvolatile liquids) and the improvement of defensive measures against them. The pace of this work increased from 1948 as the cold war began, and nations built up stocks of munitions containing nerve agents. This trend was reversed in the United Kingdom in the late 1950s when a decision was made to destroy its stocks of chemical agents. The Soviet Union and the United States were left as the principal chemical warfare powers. Other types of chemicals were examined for usefulness in war, the principal class being the psychochemicals—substances that act on the central nervous system to create hallucinations or generally disconnect the victims from reality. These included the seminatural LSD (popular with hippy cults in the 1960s) and the synthetic BZ, which at one time formed part of the U.S. chemical arsenal. (For code letters for chemical compounds, see Appendix 2.) The utility of such materials in war was always doubtful, as the effects varied from individual to individual, and perhaps with the mood of the person at the particular time of exposure. The result might be complete withdrawal from reality and passivity, or manic, uncontrollable activity. I believe the Americans have burnt all their BZ, and psychochemicals are not now thought to be useful in war.

My historical survey now approaches recent times, and the relevant technical aspects of chemical warfare and defence are given in the following chapters. The mood in the later 1960s and 1970s turned towards disarmament, with national efforts channelled through the United Nations, or bilateral negotiations occurring between the United States and the Soviet Union. The negotiations towards chemical disarmament received boosts in the late 1970s when a number of nations expressed support through the United Nations, and in the early 1980s after the United States showed support by submitting a draft treaty to the UN, presented by the then vice-president, George Bush. It was agreed that the proposed Chemical Weapon Convention would necessarily have a high technical content; therefore your attention to the next chapters will help you understand these negotiations.

References

1. Butler, A. R., and R. A. Reid. (1986). "Whence Came Chemistry?" *Chemistry in Britain* 22 (April): 311–12.
2. al-Hassan, A. Y., and D. R. Hill. (1986). *Islamic Technology.* Cambridge: Cambridge University Press.
3. Haber, L. F. (1986). *The Poisonous Cloud: Chemical Warfare in the First World War.* Oxford: Clarendon Press.
4. Foulkes, C. H. (1934). *"Gas!": The Story of the Special Brigade.* Edinburgh: Blackwood.
5. Hartcup, E. (1988). *The War of Invention: Scientific Developments, 1914–18.* London: Brassey.
6. Jones, D. P. (1983). "Chemical Warfare Research during World War I." In *Chemistry and Modern Society,* ed. J. Parascandola and J. C. Whorton, pp. 165–86. Washington, D.C.: American Chemical Society.
7. Hyde, H. M., and G. R. F. Nuttall. (1937). *Air Defence and the Civil Population.* London: Cresset Press.
8. Glover, C. W. (1938). *Civil Defence.* London: Chapman and Hall.
9. Crone, H. D. (1986). *Chemicals and Society: A Guide to the New Chemical Age.* Cambridge: Cambridge University Press.

Other books on the history of chemical warfare

R. Harris and J. Paxman (1982). *A Higher Form of Killing: The Secret Story of Gas and Germ Warfare.* London: Chatto and Windus.

S. Murphy, A. Hay, and S. Rose (1984). *No Fire, No Thunder.* London: Pluto Press.

E. M. Spiers (1986). *Chemical Warfare.* London: Macmillan.

W. Moore (1987). *Gas Attack!: Chemical Warfare 1915–18 and Afterwards.* London: Leo Cooper.

V. Utgoff (1990). *The Challenge of Chemical Weapons: An American Perspective.* London: Macmillan.

4

◁══▷

The target

Let me interrupt the history and technology of chemical warfare to introduce a topic that is often forgotten in this type of study, namely the target. The target is man – or, more exactly, man, woman and child as the degree of discrimination exercised by chemical weapons is zero. Most of the books I have read on the subject of chemical warfare describe the chemistry, the hardware, the protective devices and so forth but ignore another significant element – the sensitive and vulnerable, yet adaptable and hardy, animal that constitutes the primary target of chemical weapons. Yes, it is you and me and the rest of mankind. In this chapter I want to describe the target in brief, so that you can judge more precisely the threat that chemical weapons pose, and dismiss some dramatic yet unfounded fears.

Our perception of the human body can vary quite markedly, depending on our interest of the moment (Fig. 4.1). It may be a visual and sensual perception (A), or we may be interested in the structure of the body (B) and thus become anatomists. Perhaps we are not interested in the detailed structure of our physical form, but in how it functions as a complex but controlled machine. If we go to the lower end of organisation, chemistry may be our main interest and then we study the biochemistry of the body (C – which shows a few types of chemical reactions of the many that are necessary). The point I want to make is that when considering the body as target, we may have several different concepts of what we are examining, and we need to choose the appropriate concept of the argument we are pursuing.

A human being is a highly evolved creature in a structural sense, yet not uniquely so. Nor are humans the peak of only one line of development, leading from primitive vertebrates through reptiles to mammals, for there are quite different lines of animal evolution that have produced structural com-

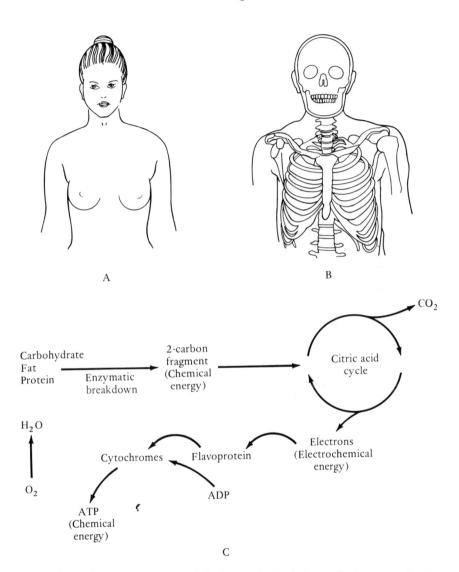

Figure 4.1. Differing perceptions of the human body. A. An aesthetic perception by an artist. B. The way an anatomist might think of a human figure. C. A biochemist's view, in terms of chemical pathways: here the transformation of chemical energy in foodstuffs to more accessible energy in ATP. Don't worry about what the symbols actually mean.

Table 4.1 *Levels of organization within the body,*
and the associated study

Level	Associated study
Whole body	Anatomy, physiology, pharmacology
Organs	—
Tissues	Histology
Cells	Cellular physiology
Organelles (mitochondria, nuclei, etc.)	Biochemistry, molecular biology
Chemical molecules	Biochemistry, molecular biology

plexity. An example is the evolution of the cephalopod molluscs (squids, octopuses) from primitive ancestors, whose cousins are today unremarkable slugs and snails. We also make a mistake if we think that structural complexity is the only criterion by which we should assess development. We can use quite a different standard and assess the ability of the organism to process chemicals – what variety of chemicals it can make use of for providing energy and structural material for growth. By this standard, many uninteresting bacteria suddenly become much more complex than humans, because they often have a more flexible chemical potential than the human body. The point of this paragraph is to show that we have to consider the human species both as a system of great structural complexity, which requires elaborate control systems, and also as a chemical factory with the ability to deal with many chemicals, but not all.

The structure of the human body can be considered in terms of levels of organisation, which start at the chemical level (that of molecules) through to the level of cells, then tissues, then organs and finally to the whole body (Table 4.1). The specific sciences associated with the various levels are also given in the table. These refer to different aspects of the organisation, for histology is a study of the structure of tissues, pharmacology that of some aspects of the controlling mechanisms of the body (physiology covers others) and molecular biology is reserved for the study of very large molecules such as nucleic acids and proteins. The structural aspects are quite obvious, but what is all important is that this structural whole must be kept as a functional unity by a complex of control measures that hold each level together, and integrate it with the levels above and below. So it is the control systems that are critically important, yet structure interacts with them to allow the control to occur. Consider the bottom level, that of the chemistry. This could not be controlled if all the chemicals were present in solution in one bucket of a cell. It is a very complex structure of large molecules and

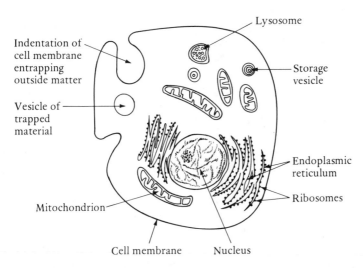

Figure 4.2. Structures within an animal cell. This highly diagrammatic representation nevertheless shows the larger organelles visible under the light microscope and the finer structure visible to the electron microscopist. The structure enables the cell processes to be ordered – for example, the production of energy from fat or carbohydrate oxidation is performed by ordered enzyme systems on the inner walls of the mitochondria. Similarly, ribonucleic acid (RNA) on the ribosomes codes the sequential assembly of protein molecules. Lysosomes contain packets of enzymes, which by contrast function in the breakdown of proteins, nucleic acids and carbohydrates.

membranes that allows chemical reactions to be kept in sequence, regulated by influences other than those of basic physical chemistry and adapted to suit the requirements of the body at a remote distance. Structure and control are thus two interdependent entities.

Structure begins to have importance at the molecular level, for it is the three-dimensional structures of large proteins, the enzymes, that enable them chemically to transform much smaller molecules of very particular types. The structure of the enzyme is necessary both to allow it to select the particular molecules on which it has to act, and also to cause the chemical activation that will, for example, convert sucrose into glucose and fructose. Similarly, the synthesis of specific protein is dependent on the structure of nucleic acids, which carry the code for the correct sequencing of aminoacids to form those particular proteins.

Within cells are structures (Fig. 4.2) that, though formed from many large molecules, are very small. One example is the mitochondria, which are the bodies within the cells that convert the energy in the breakdown products of glucose into the energy-rich phosphate compound known as ATP, the near-universal currency of energy in living organisms. Cells themselves are invisible to the naked eye, but numbers of like cells form tissues, which in turn appear in various organs of the body. Muscle, for example, has three types, which appear in muscle as we know it (e.g., biceps), in the heart and in smaller aggregates in the intestinal wall and elsewhere. The organs of the body are familiar to us, perhaps more often recognised when they malfunction rather than when they quietly do their job. The heart is a pump, the liver a chemical factory, the kidneys remove waste, the larynx generates sound and so on. Finally, the whole assortment of cells, tissues and organs is tied into the one functional whole that is your body.

The control of this whole apparatus is by chemical means. Certainly, as you have no doubt commented, the nervous system does convey electrical signals along the nerves to remote parts of the body, but when the signal arrives, it is converted into a chemical one. Each electrical pulse becomes a package of molecules, which are released at the nerve ending and diffuse locally in the adjacent tissue to cause a particular effect. Also, in the dense electrical circuitry of the brain, the intensity and routing of signals is determined by many different chemical influences. The nervous system serves to convey chemical signals fairly precisely to defined points. More general and diffuse instructions are given simply by the local release of a chemical messenger in a tissue or organ. Messages to go throughout the whole body are carried by chemicals released by one organ into the bloodstream, the familiar hormones. Thus the release of insulin from the pancreas carries a general message to all cells that the release of glucose into the bloodstream is to be reduced, or the production of male hormone from the testes promotes hair growth on the face and enlarging of the larynx in the adolescent child.

Let's put the chemistry aside for the moment and consider what relationship this discussion of structure and control has to chemicals used in warfare. If you think about it for a time, it is clear that very toxic chemicals must act on control systems, rather than just nonspecifically consume the victim's body. After all, it will take a fair quantity of something even as corrosive as sulphuric acid to damage a person fatally, whereas the chemical warfare agents are toxic in the order of milligrams per person (for VX [see Appendix 2] a fatal dose could be 5 mg or a ratio by weight to the victim's weight of 1 to 14 million). In fact, the most toxic warfare agents do act on control systems, but some quite dangerous chemicals have a simple blocking action. Thus hydrogen cyanide and carbon monoxide both, in different ways, interfere with the carriage of oxygen to the tissues and cells for use

Table 4.2 *A classification of the means by which toxic chemicals act*

Action	Effect and examples
Corrosive	Simple, nonspecific destruction of tissue: sulphuric acid, strong alkali
Blocking	Stop a chemical process in the body (metabolic inhibition) or interfere with a structural component: hydrogen cyanide, carbon monoxide, heavy metals (mercury, lead)
Disruptive of a control system	
At the cellular level	Interfere with a system that controls a function such as protein synthesis: mustard gas
At the organ or body level	Interfere with a system that exercises control throughout the body: nerve agents, opiates

by the mitochondria. The nerve agents and mustard, however, interfere with control systems, the first at the whole-body level, the second at the level of protein synthesis from the coded nucleic acids. Nerve agents interfere with the chemical signal that carries a message across the gap between two nerve cells (the synapse) or between the nerve cell's elongated arm and the membrane of muscle cells. The signal is not stopped by nerve agents but is repeated excessively, so that the muscle, for example, contracts much more than is necessary and may end in semipermanent convulsive contraction (tetanus). The effect is to disrupt message transmission, both within the brain and down the long fibres of the nerves (axons) to the rest of the body. The initial action of mustard is to alter chemically the nucleic acids (DNA and RNA) in cells, and the final outcome is the death of the cells. What happens in between is still being investigated; one hypothesis is that disruption of control mechanisms in the cells caused by modification of the nucleic acids leads to the release of enzymes that destroy the protein and lipid architecture of the cells, leading to cell death. Other hypotheses are possible, but the effect is essentially the loss of ordered control mechanisms. Mustard and nerve agents lie at opposite ends of the organisational spectrum, mustard attacking all cells, whereas nerve agents attack one specific protein type at the critical points of a system that extends throughout the body. All toxic effects, however, are at the point of action chemical events, which lead to cell-specific or organ-specific changes.

The different types of actions displayed by various toxic chemicals as previously discussed can be summarised simply as in Table 4.2. The distinction between corrosive and the others is quite clear, as is that between effects at cellular level and body level. One can argue, however, about some aspects of blocking versus disruption at the cellular level because mustard, for example, can be regarded as blocking protein synthesis, although the block occurs on an information-carrying and thus controlling molecule. Nevertheless, Table 4.2 gives an idea of the distinctions that can be made.

A very small dose of a nerve agent circulates through the bloodstream, after being absorbed through the lungs or skin, and then attaches to the comparatively few molecules at which it has most effect. Obviously, therefore, there must be a great deal of specificity in this attachment, for otherwise the nerve agent would rapidly be mopped up by the billions of other molecules in the body. Mustard also has specificity for certain attachment points on nucleic acids, and we can see that specificity of attachment and action is a characteristic of very toxic chemicals.

The organisational complexity of the body thus creates opportunities for very specific chemicals to exercise a deleterious effect, an effect that is enhanced when control systems are the target. How does the chemical capability of the body influence toxic effects?

Although perhaps not so chemically flexible as bacterial cells, some cells of the human body do have great ability to adapt themselves to process foreign chemicals. The liver, the chief site of this activity, is well served by blood vessels so that any material entering the bloodstream from the lungs, intestine or skin is carried to the liver in a matter of seconds. A great many enzymes in the liver can break down or modify foreign chemicals, which are then passed into the intestine in the bile or modified to a form that the kidneys can rapidly excrete in the urine. Furthermore, the presence of foreign chemicals can induce in the liver the formation of additional enzymes that will break them down. Thus the liver seems to retain a few molecules of a wide range of enzymes, or perhaps only the coded patterns for them, which can be switched into full production by the presence of the substrate, that is, the foreign chemical, the material acted on by the enzyme.

The chemical activity of the body, as presented here is almost totally protective, but note that chemical mechanisms are susceptible to poisoning, as mentioned earlier for the action of hydrogen cyanide. In fact, poisoning is a chemical event. What I am trying to convey is that it is much more effective when aimed at a control system with a structural aspect. Chemical activity, as well as breaking down a poison, can also in rare cases enhance the toxic effect, but this is so unusual as to be of little significance. In effect, the body has a chemical organisation to protect it.

An example is that of the fate of the nerve agents, the organophosphorus esters. These can be broken down by a family of enzymes, the phosphatases, which are present in the blood and most tissues. Further, the nerve agent

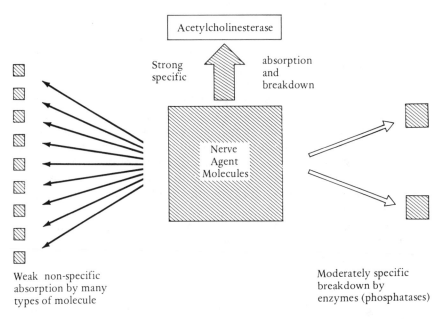

Figure 4.3. A representation of the competing influences that determine the loss of nerve agents from the bloodstream.

circulating through the body in the blood will be absorbed by many chemicals in a more or less specific manner and removed from circulation. The net result is that the nerve agent does not survive in the blood for more than 30 to 150 seconds. Does this not mean that something which is broken-down or absorbed so quickly is a poor poison? It might if it were not for the extraordinary specificity of the target enzyme, the acetylcholinesterase, for the nerve agent. The scavenging and binding action of this enzyme is so strong that the nerve agent is bound to the enzyme after one or two passes through the bloodstream (Fig. 4.3). In the competition between various binding sites, the acetylcholinesterase wins (or loses) every time. If this balance was altered, the effect of the nerve agent might be reduced, and this procedure has been tried. Efforts have been made to induce more phosphatases in the body before exposure to nerve agents, but these experiments in animals have not been too successful. The idea is that people who might be exposed to nerve agents would be given nontoxic organophosphorus compounds, which would induce phosphatases in their bodies at much higher than normal levels.

Nerve agents are small molecules, too small to induce the formation of antibodies in the human body. Attempts to make them better antigens by binding them onto larger molecules have been made but have not resulted in a practical protective system. Protein toxins will induce antibodies, however, and therefore it is possible to protect people against such toxins by pretreatment with small doses of the toxin, or toxinlike molecules.

You will understand from the foregoing that a description of the target in terms of both its strengths and weaknesses helps to explain how chemical warfare agents exert their effect and allows an assessment of the vulnerability of the target. The gross simplification that I have given here is not intended to give you detail, only a conceptual grasp of the topic.

The target, the body, has other practical lines of defence, which I consider more fully in Chapter 6, because they relate closely to the artificial defences that we can adopt.

It is important to remember that very few other methods of waging war are as specific to the target as is chemical warfare. It is quite possible to kill all human beings in an area with a volatile nerve agent without damaging plants or any material structure at all. Many animals would also die, but the more primitive in organisation would live. By contrast, modern "conventional" war has become a destruction of matériel, human death being incident upon the demolition of a structure or the generation of flying debris. The newsreel and television images of destruction are infinite; from my childhood come Caen and Monte Cassino, because of the architectural inheritance that was lost. The only weapons that approach the chemical in selectivity are the neutron bomb and the biological weapons, which are the most specific of all but unpredictable in some respects.

Some understanding of the target for these specific weapons was therefore necessary, but now we can consider the technical aspects of the weapon and the means of defence against it. I have deferred one property of the target, how it protects itself from foreign chemicals, until Chapter 6, because it is best discussed in relation to protection by artificial means.

5

◁══▷

Modern chemical warfare: offence

What is a chemical weapon? Is it a particular chemical that is so toxic and destructive of human life that it can have no other use than in war? Or is it a device, containing a toxic chemical, that is designed to release its contents over an enemy and achieve the best possible distribution of its poisonous payload? My belief is that the latter is the clearest definition. The former is only half the necessary definition, for there are many toxic chemicals that have a possible use in war (indeed have been used in war), yet have legitimate uses for peaceful purposes. Phosgene and hydrogen cyanide are good examples; each is produced and used in industry by the thousands of tonnes as basic components for synthetic chemistry. The inclusion of a chemical into a delivery system is an unambiguous statement of its final purpose. One may have doubts about the ultimate use of phosgene by a nation's industry, if that nation has little chemical output, but one has no doubts at all about the intended use of phosgene in an aircraft bomb, filled with a fuse and a burster mechanism. A chemical weapon thus has two elements: a highly toxic chemical fill and a delivery system to distribute it onto the target.

There are several ways of classifying the various chemicals that can be used in weapons; my method in Table 5.1 is probably most suited for our purpose. This is based on the likely use of the chemicals in war, derived from past experience. (For code letters for chemical compounds, see Appendix 2.)

Chemical stocks declared by the United States and the former Soviet Union have included sulphur mustard (H) and nerve agents. The Iraqi arsenal developed in the 1980s also consisted of mustard and nerve agents. As recounted in the previous chapter, the post–World War II choice was mustard and nerve agent, so that those two classes of chemicals are the

Table 5.1 *A classification of chemicals for chemical weapons,*
based on the likely use as judged by past experience

Class	Examples
1. Chemicals recently used or known to be in stocks	Mustard (H, HD, HT) Nerve agents (GA, GB, GD, VX)
2. Industrial chemicals, used in World War I	Chlorine, phosgene, hydrogen cyanide
3. Chemicals once weaponised or regarded as possible weapon fills	Nitrogen mustards, Lewisite (perhaps should be in class 1), other arsenicals (Adamsite, etc.), many other compounds
4. Toxins – poisonous compounds originally derived from living organisms	Botulinum toxin, ricin, others have been investigated
5. Possible new developments – chemicals that may have been evaluated for war	PFIB (see text) mycotoxins, pharmacological controlling drugs, BZ, many others

most likely to be selected for future use, if any. There have been good technical reasons for selecting them, which include high toxicity and suitability for use in the battlefield.

The second of my classes included World War I chemicals, but one that cannot be neglected now. Chlorine is unlikely to be used, as it is not very toxic compared with the other two. Phosgene is probably more dangerous than hydrogen cyanide; it was put into aircraft bombs in World War II, and it would be militarily effective. The fact that all three of these chemicals are widely used in industry makes their control very difficult.

A vast number of chemicals were examined for their usefulness in weapons during and after World War I. Perhaps the greatest single class was that of the compounds of arsenic, including Lewisite (which causes burns like mustard) and Adamsite (a solid that induces nausea and uncontrollable vomiting). Lewisite found a use as in 1:1 mixture with mustard (H), as this lowered the freezing point of the mustard, a useful result in cold climates. The mixture (known as HL) found its way into Russian and Japanese arsenals. Lewisite today is probably not a likely choice. The nitrogen mustards (HN) are similar in effect to sulphur mustard as they both transfer part of their molecular structure to nucleic acids and other essential components of cells, causing cell death. One nitrogen mustard, under the name of mustine, is still used as an anticancer drug; the hope is that it kills the rapidly growing tumour cells before it affects the normal ones. For various reasons, none of the class 3 compounds are now chemicals of first choice for war.

My class 4, toxins, introduces all sorts of difficulties in defining them. First, they are explicitly covered by the Bacteriological (Biological) and

Toxin Weapons Convention of 1972, yet are chemicals rather than biological warfare agents. The latter are infectious organisms, capable of reproducing themselves and growing rapidly in numbers once they infect a suitable host (usually human, when the intent is war). Chemicals are artificial and cannot increase by themselves; they are nonliving matter. In between are toxins, neither capable of reproduction nor able to be made by humans, until recent times. Modern synthetic chemistry has enabled us to make more and more complex molecules, so that the natural products of living organisms can often be made synthetically. We can now make the simpler toxins and, in principle, could make the more complex. Therefore the former distinction of toxins as a separate class is losing its meaning, for they now become chemicals like phosgene or nerve agents. A future chemical convention will have to mention toxins, in order to define the relationship between it and the 1972 Convention on Biological Warfare.

The other controversial point about toxins is whether they have any practical utility in weapons. Certainly they are extremely toxic, many more so than the nerve agents, but they are all solids and this physical form, as we will see later, is not the most useful for a chemical in a weapon. The poisonous principle of the castor oil bean, ricin, was used in the Bulgarian Umbrella Case – no, not a Sherlock Holmes story. A Bulgarian émigré in London, Georgi Markov, was shot in the leg with a minute sphere of a platinum-iridium alloy from a modified umbrella operated by a Bulgarian agent. The pellet (1.53 mm in diameter) was drilled through and in the cavity was 0.5 mg of ricin. Markov died some time later, of symptoms suggestive of ricin poisoning. By a miracle, the tiny pellet was found during the postmortem examination. This story illustrates two points about toxins: one, the remarkable toxicity they can command (for ricin, the intravenous LD_{50} is about 10 micrograms per kilogram); two, the elaborate means necessary to administer them in order to take advantage of this toxicity.

My fifth and last class (Table 5.1) includes a miscellany of possibilities, which I will discuss further at the end of the chapter when I consider the future.

Some of you will say I have left out an important class of chemical, the tear gases or riot-control agents. Others will say that I have left out herbicides that could be used to destroy crops and starve an enemy. Both of these types of chemicals have, however, a very important distinction from the classes of chemical warfare agents I have considered. They are carefully tested to ensure they have no long-term toxic effect on humans, whereas the warfare agents are chosen for the opposite reason, of having extreme human toxicity.

Considered within the concept of toxicity only, we are dealing with two opposites. Herbicides have low toxicity to humans and are being further improved in this respect. The long, unresolved argument about the human effects of the use of Agent Orange in Vietnam illustrates the distinction between this and a human-directed chemical; the argument is unresolved because the effects were nonexistent or slight. There is no doubt about the effects of a chemical warfare agent. Riot-control agents are extremely toxic to humans in the short term, but have little or no effect in the long term. Thus exposure to CS produces instant, overwhelming discomfort, yet its toxicity in the long term is very low.

We have to create a new group of chemicals, that is, chemicals of use in war, as opposed to chemical warfare agents. The former have low, long-term toxicity to humans, the latter have extremely high toxicity. The former are often not directed towards humans, whereas the latter are always so directed. The riot-control agents when used in war can be regarded as warfare agents but retain a permitted use in domestic riot control. Some people might want the tear gases banned from the latter role also, but there is a humanistic reason to retain them. If the alternatives to quell a riot are the use of tear gas or of wooden truncheons with which to hit people over the head, then the former is preferable. Tear gas has no long-term effect (if used sensibly in the open), whereas permanent brain damage can result from blows to the head. Riot-control agents thus need a special niche in a chemical warfare convention. (See Chapter 9.)

To revert to the main theme, the typical chemical weapons of recent times are the nerve agents and sulphur mustard. The nerve agents all have the same toxic mechanism, which is to disrupt the passage of nerve impulses between nerve and nerve and between nerve and muscle. The immediate effect is that the impulses are not switched off but continue excessively so that normal muscular movement becomes convulsive contractions. The effects of disturbance within the brain also are transmitted throughout the body. Death is a result of a number of effects, the most obvious being asphyxiation as the regular ventilation of the chest ceases. Nerve agents principally differ in their physical properties; all are liquids, but they range in volatility from being as volatile as water, as in GB, through intermediate ranges, as in GD and GA, to being nearly nonvolatile, as in VX.

Nerve agents interfere with the sophisticated control mechanisms of the body, whereas mustard, as I have indicated, kills indiscriminately any cell with which it comes in contact. The skin and lungs most obviously suffer, as they are the first tissues in the body to be reached by liquid or gaseous mustard. Mustard on the skin causes blisters or vesicles, some time after it has killed the cells, so that it is classed as a vesicant "gas." It is not, in fact, a gas as it is a rather nonvolatile liquid, but more volatile than VX and just enough so that the vapour is a very toxic hazard. The liquid penetrates into many materials, is chemically fairly stable and dissolves very slowly in water. Mustard therefore hangs around any area in which it is dispersed for long periods and is a general nasty nuisance. Fisherman in the North Sea are currently hauling up jellied gobs of mustard from munitions jettisoned forty-five years ago after World War II.

Why jellied gobs? Because liquid chemical agents such as mustard or VX, when disseminated above a target by an explosive change, get blown into such fine drops that they drift in any breeze and do not fall on the target. When thickened to a viscous mass, the droplets remain larger and fall to ground. In World War II, some types of mustard munitions were first thickened with chlorinated rubber, then with ground up Plexiglas from the canopies of crashed Hurricanes and Spitfires.

You will have noticed the code letters I have sometimes used for chemical agents – H, GA, VX. The H for mustard is said to have originated in World War I as HS, an abbreviation for "hot stuff." Mustard also has been coded as Y for Yperite, a French term originally. In Appendix 2 I have given a guide to the bewildering terminology of warfare agents. Two common errors need correction. HL is not mustard prepared by the Levinstein process: it is the 1:1 mixture of mustard and Lewisite I described previously. The term HT refers to mustard containing 40 percent of a related compound, T, which is a double molecule of mustard. HT does not mean thickened H, which was designated by V for viscous. Thus HT (V) was thickened HT.

Why are mustard and nerve agents the chemicals of choice for war? It is not simply because of their toxicity. There are substances much more destructive of cells than mustard (e.g., the T-2 mycotoxin from moulds) or more powerfully effective on the nervous system than GD (some of the carbamate esters). The mycotoxins and carbamates, however, are solids and therefore are less readily dispersed and do not enter the human body as easily as do liquids (see Chapter 6 for further discussion). Mustard vapour can be inhaled, resulting in destruction of the lungs, and the liquid spreads on and penetrates skin very quickly. Solid mycotoxins are not easily inhaled, unless maintained as a fine aerosol, and the power penetrates still less readily. Similar considerations apply to carbamates as opposed to nerve agents. In other words, the physical form of the chemical determines the ease of access to the human body, which greatly influences its practical toxicity. The final toxic hazard of a chemical is also determined (Fig. 5.1) by events within the body, such as speed of elimination and of breakdown (toxicokinetics). For a fuller treatment of the factors that determine overall hazard from a toxic chemical see my earlier review.[1]

Other practical matters influence the choice of chemicals for use as warfare agents: ease of manufacture, cost, stability when stored in a weapon,

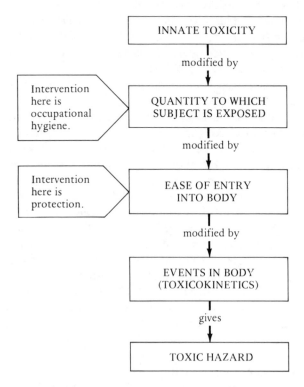

Figure 5.1. Successive modification of the innate toxicity of a substance to result in the toxic hazard. In the military context, occupational hygiene is partly replaced by the practice of contamination control.

difficulty of providing protection, difficulty of treating casualties after exposure. These and other factors all favour nerve agents and mustard. They are simple molecules, readily made in bulk at low cost, are stable in storage and have an effect immediately on the victim. (The onset of obvious symptoms of mustard poisoning is slow, but an irreversible train of toxic events commences a minute or so after contact with living tissue.) What more could be desired? In particular, the nerve agents can supply an infinite range of compounds with physical properties tailored to the requirements, from volatile GB as a threat to the lungs to VX as a penetrant of skin. A popular concern about possible new agents with greater and greater toxicity arises, I believe, from an inability to appreciate how toxic the nerve agents are. One drop of VX left on your skin can kill you; one breath of air containing GB vapour can also kill you. There is really no need for anything "better," and these compounds are the ideal weapons of chemical war.

But let me make one correction, to make the last statement more exact: Mustard and nerve agents are the ideal weapons of chemical war when allied with effective delivery systems. It can be argued that the most threatening advance in chemical warfare in the past half century has been the development not only of nerve agents but also of the means to carry them to the target. The technology available in the 1940s of artillery shell and aircraft bomb or spray were not particularly effective, for the shell had to be mainly steel to withstand the pressure in the barrel of the gun and therefore its payload of mustard was low, and the aircraft engaged in spraying had to fly low and thus be vulnerable to ground fire. I suppose the aircraft bombs would be quite effective in blanketing an area with chemical agent; during World War II, the United Kingdom's light-case 250-pound bombs carried about 120 pounds (55 kg) of mustard or phosgene, enough to contaminate 11,000 square metres to 5 g/m^2, or one bomb needed per hectare. This bomb in modern adaptation became an Iraqi weapon of the 1980s; a case originally intended for white phosphorus was filled with mustard and then burst open by the detonation of a central tube filled with high explosive (Fig. 5.2).

The requirements for a delivery vehicle are a high ratio of chemical payload to structure (i.e., the munition wall) and the ability to cover an area of ground rapidly in as short a time as possible, in order to take maximum advantage of surprise. The multibarrelled rocket launchers, modern developments of the Katyushka rockets favoured by the Russians in World War II, meet those requirements because a salvo of forty rockets can be fired in thirty seconds. Other modern developments are large missiles, ballistic or guided, which could carry chemical payloads of tens to hundreds of kilograms to targets relatively close or on the next continent. Whether intercontinental missiles have been constructed to take chemical payloads is unknown, but the possibility is there. Modern technology certainly offers the means of making chemical war more deadly, and the lessons are learnt, for the Iraqis also have used rockets with nerve agent payloads (Table 5.2).

In Table 5.3, I have compared the two delivery systems mentioned, the British World War II aerial mustard bomb and the modern Soviet BM-21 rocket launcher. The bomb is simple and reasonably effective, but warning of attack is given by the approach of the enemy aircraft and the dispersion of the agent payload is poor; because the payload all comes from one point source, the ideal area of dispersion of 1 hectare will not be achieved. The BM-21 is, by contrast, almost a perfect delivery system: The launcher system mounted on one truck can deliver a fairly even coverage over 2

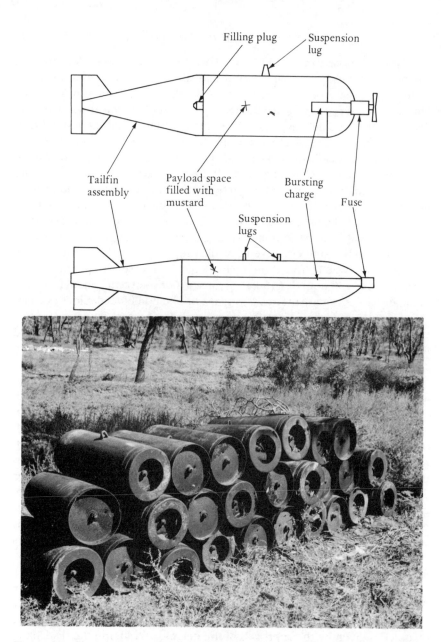

Figure 5.2. Chemical aircraft bombs. *Top:* Diagrams of World War II 250-pound bomb (upper) and Iraqi 135-kg bomb from the 1980s. The similarity in construction is obvious. *Bottom:* Empty World War II 250-pound bombs, originally filled with phosgene. The view is from the tail ends, where the fins were fitted. The filling plugs are visible. From the records of the Materials Research Laboratory, Melbourne. The details of the Iraqi bomb were supplied by P. Dunn.

Table 5.2 *Chemical weapons declared by the United States and the former Soviet Union to be in their possession, and those known to be possessed by Iraq in 1991*

Country	Weapons
Soviet Union	Missiles (VX and thickened VX), artillery shells (GB, VX and thickened L), rockets (GB and VX), aircraft bombs (GB and HL), spray tanks (thickened GD and HL), grenades (CS)
United States	Artillery shells (GB, VX and H), aircraft bombs (GB), cluster bombs (BZ), rockets (GB and VX), mortar bombs (H), spray tanks (GB and VX), mines (VX); binary shells (GB) and binary bombs (VX) and rockets also in advanced development
Iraq	Artillery shells (H), rockets (GB/GF mix), aircraft bombs (H and GB/GF mix), missiles (GB/GF mix), mortar bombs (CS); unmixed GA and GB also produced prior to 1988

Table 5.3 *The performance of two chemical delivery systems, the British World War II 250-pound bomb and the modern Soviet BM-21 rocket launcher*

	250-pound bomb	BM-21
Type of system	Single, light-case bomb	40-tube multiple-rocket launcher
Chemical agent payload	55 kg	$40 \times 3 = 120$ kg
Time to fire munitions	[Single unit]	< 30 seconds
Area coverage to density of 5 g/m^2 (ideal dispersion)	1 hectare	2 hectares
Effectiveness of dispersion	Poor (from 1 point)	Good (from 40 points)
Warning signs	Approach of enemy aircraft	None

hectares within thirty seconds in complete surprise. Against unmasked troops, the casualty rate from a GB attack would approach 100 percent. Conversely – such is the paradox of chemical defence – against masked troops the attack would be totally ineffective. (See Chapter 6.)

The question of the useful payload of various delivery systems is illustrated in Table 5.4, in which I have expressed the chemical content as a percentage of the weight of the projectile and of the combined weights of the projectile and propellant. The projectile weight is not a useful basis of comparison between classes of missiles, because the artillery shell leaves its propellant in the gun barrel, whereas the initial launch weight of a rocket includes the rocket motor and fuel. The combined weights of projectile and

Table 5.4 *The toxic chemical content of various artillery missiles as a percentage of the total weights of the projectiles and of the complete systems*

	Chemical content as percentage of	
	Projectile weight	Total projectile and propellant system
Artillery shells	8	6.5
Rockets (multiple launch)	15	7
Missiles – e.g., FROG-7, SCUD-B	[53]	8.6

Notes: Because the chemical contents of different types of each projectile vary, the difference between projectiles in the right-hand column (e.g., shells versus rockets) is not significant. Also, The projectile weight percentage given for the missiles is in fact that of the warhead alone.

necessary propellant form a better basis and one that is more meaningful in terms of supply and logistics – how much material you have to move to the artillery battery in order to deliver so much chemical agent.

Cruder technology is also available for close-in fighting and for terrorist use; chemical mines, Livens projectors, smoke candles and frangible grenades are all old but simple devices readily improvised. Frangible grenades of Japanese origin were captured in World War II by Australian forces and brought back to our laboratory for examination. The grenade was made of a thin-walled glass sphere, perhaps 125 mm in diameter, containing liquid hydrogen cyanide or phosgene and packed in a cardboard box. It was intended to be thrown into a pillbox or bunker to release suddenly a cloud of gas; it would be useless in the open air as the gas would be diluted too quickly to be effective. Many other weird and dangerous items were devised in World War II, among them an idea of using thickened liquid hydrogen cyanide in a dispenser resembling a flame thrower and supplied from a backpack. The idea was to project a lance of hydrogen cyanide into the driver's window of an oncoming tank, or into a bunker. It was abandoned, no doubt because it was rather more dangerous to the user than to anyone else.

There is no need to give more detail on delivery methods, other than to summarise some types (Table 5.5). Enough has been written to make it clear that the delivery device is an essential part of the chemical munition, which must be recognised as the combination of toxic payload and dispersing mechanism.

Table 5.5 *A summary list of devices for delivering
or spreading toxic chemicals*

Type	Description
Gas cloud	Release of gas from cylinders, relying on wind to carry to enemy; obsolete
Livens projector	A crude form of mortar, best described as a drum of chemical fired from a drainpipe
Smoke candle	A pyrotechnic device producing heat that evaporates or sublimes a solid agent
Artillery and mortar shells	See text
Aircraft sprays	Efficient way of covering an area, but aircraft is vulnerable to groundfire
Aircraft bombs	See text
Rockets and large missiles	See text
Grenades	Useful for street fighting
Mines	Mine with chemical fill and bursting charge; useful for ambush and terrain denial

In a more general way it can also be seen from the foregoing that chemical weapons are attractive items for an aggressor wishing to gain a military advantage. However, they do have disadvantages in use, so that a country that blunders into using them may come to regret such action later.

The first disadvantage is that they can be rendered ineffective by good protective measures, as we will see in the next chapter. By "good" I mean really good, but protection is certainly possible, and a well-prepared defence negates the usefulness of the weapons. As a result, chemical weapons are most decisive when used to surprise the opponent and catch troops before they can don protective gear. Second, chemicals do not recognise friend or foe. Escaping from a factory or munition-filling plant, the chemical will cause friendly casualties. Similarly, on a battlefield, the attacker will need to wear protective gear as well as the defenders. It has been argued that the attacker can fire chemical shells from a distance or drive through the contaminated area in a sealed vehicle (see Chapter 6) and thus not need protection. This is a very naive argument, for if I was firing chemical shells from a 155-mm howitzer at an enemy 15 km away (not my choice of occupation), I would be wearing full protective equipment to meet two eventualities: (1) that my gun would misfire and burst a shell in the barrel and (2) that the enemy would land a conventional shell among my stock of chemical ones. As regards chemical weapon manufacture, the workers in

World War II mustard factories have provided a harvest of epidemiological data on the effects of exposure to mustard; Japanese and British[2] data are available. Meddling with toxic chemicals inevitably brings a risk.

Third, once the chemicals are made and loaded into munitions, they are embarrassingly difficult to get rid of. One obvious way is to fire them at the enemy, but even then the successful army has a problem in cleaning up the contamination left behind, if anything more persistent than GB or hydrogen cyanide has been used. Admittedly, once the correct procedures are learnt, this may not be any more difficult than clearing a minefield (not an easy job). More problems occur if the munitions are not used but left in magazines to corrode slowly away. The U.S. and the former Soviet chemical arsenals have proved a problem in this respect, and the chemical stockpile of the Iraqis, now partially destroyed by U.S. bombing, poses the additional problem of munitions ruptured and scattered by that bombing. World War II chemical munitions still turn up in unexpected places; I referred earlier to the thickened mustard netted by fishermen in the North Sea, and mustard munitions or containers are found occasionally at the sites of old munition dumps. These are often mistaken for conventional munitions, until the charge set by the disposal unit results in a shower of liquid rather than an explosion.

There are disadvantages in the use of chemical weapons, but also advantages in the eyes of the irresponsible few who wish to gain a short-term military advantage, especially those with no permanent responsibilities. Another consequence of the attractiveness of chemical weapons is the need for all nations to maintain defensive measures, which largely negate the weapons' advantage. (See the next chapter.)

What of the future? If there is no effective law on chemical weaponry, it is certain that attempts will be made to develop "better" weapons, and this can follow several lines of perceived improvement. The United States has tried to make the weapons inherently safer in storage and handling by developing the binary munition concept. Two precursor chemicals are mixed when the munition is fired or released, to produce the very toxic product during the flight to the target. The precursors are toxic themselves, but not nearly as much as the product. Another line of development is to design chemicals that will have specific properties enabling them to defeat the current protective equipment. Because the most rapid route of entry to the body for chemicals is through the lungs, the respirator (gas mask) has the most stringent protective requirement, and therefore is an attractive target for attempts to defeat protection by means of newly synthesised chemicals. The current developments in biotechnology mean that the synthesis of toxins on a large scale will be easier or cheaper, so that their use in warfare will be reexamined; extremely high toxicity may compensate for unfavourable physical properties.

However, munition stockpiles have tended to be conservative, because of the logistical difficulties of getting rid of old ones. If a new chemical is developed in a laboratory, is it worthwhile adding it to the stockpile if the cost is either additional storage space or destruction of the old stock? The answer is to be very cautious about new additions.

This brief summary of the aggressive use of chemicals has necessarily been a simplification. We should be aware that quite different threats and consequences could arise from the use of chemicals other than the nerve agents and mustard. There are many other classes of toxic chemicals that could be used; the proportion of these that have the right properties to be militarily useful will be low but will still provide many possibilities. However, defence against chemicals is possible, as will be shown soon.

References

1. Crone, H. D. (1986). *Chemicals and Society: A Guide to the New Chemical Age.* Cambridge: Cambridge University Press.
2. Easton, D. F., J. Peto, and R. Doll. (1988). "Cancers of the Respiratory Tract in Mustard Gas Workers." *British Journal of Industrial Medicine* 45:652–9.

6

◁══════════════════════════════════════▷

Modern chemical warfare: defence

It is evident from the preceding chapter that chemical attack is an area attack; it is not precisely targetted in the sense that a bullet is. Because it would be most effectively employed against dense concentrations of persons, cities are prime targets. The recognition of the vulnerability of civilian targets to chemical attack dates from before World War II and accounts for the extensive precautions against chemicals that were taken just prior to that war. In Britain, gas masks were available for the whole population, and personnel of the ARP (Air Raid Precautions) were issued eye shields and antigas ointment as further defences against mustard gas, as recounted in Chapter 3. Defence against chemical weapons therefore requires much more than equipping the military forces, for every man, woman or child is a potential target.

Because these weapons are distributed over an area, they are to an extent wasteful, which helps defence to some degree as most of the chemical will not find a target. One ridiculous but common fallacy is to equate the contents of a chemical munition with a stated number of lethal doses, based on the dosage appropriate for intravenous injection. Such injection would be a very economical method of conducting a war, but it is not really practical on a battlefield.

A number of resources and strategies can be used to protect persons (Fig. 6.1):

1. The human body has mechanisms to protect itself from toxic chemicals in the environment, and an understanding of these is necessary for good defence.
2. Individual persons can be protected by gas masks, suits, gloves, and so forth.
3. A group of people may be protected by a purpose-designed structure: collective protection.

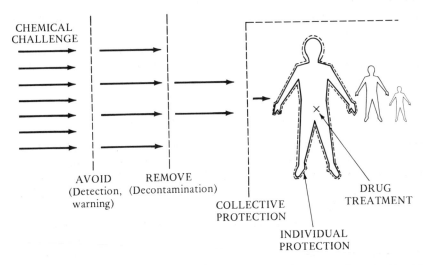

Figure 6.1. The sequence of measures that can protect the person from toxic warfare chemicals.

4. Early warning and detection of an attack will enable the targetted persons to take avoiding action and put on protective gear. Similarly, if a contamination monitoring device is available, it is possible for persons to avoid contaminated areas or highly contaminated equipment.
5. Toxic chemicals can be removed from, or destroyed on, equipment and terrain that they contaminate. This process of decontamination is useful after an attack with those chemical agents that persist as threats in the environment.
6. Medical intervention at an early stage of poisoning can markedly reduce the toxic effects of some, but not all, chemical warfare agents (therapy). It is also possible to give drugs that will protect a person against the consequences of exposure to nerve agents at some time later (prophylaxis).

Each of these processes by itself will provide little defence against chemicals, but all six together are capable of providing a high degree of protection. That degree of protection depends upon what you judge it against. It is a popular fallacy to believe that it is more difficult to protect against chemicals than against conventional weaponry. But consider a bullet aimed at your chest. It is impractical to provide you with armour that will withstand the bullet, so you will die. Then consider a lethal through-the-skin dose of VX falling on you, a drop of say 5 mg. It is perfectly practical to protect you against this or a thousand times that amount, so you will live.

Good equipment and training will provide protection against large quantities of toxic chemicals, but the equipment must be faultless and the training unimpeachable. Therein lies the weakness of chemical defence.

We can now discuss the various means of protection in a little more detail, but the matters are complex so that what follows are sketches of the technical methods.

Natural protection of the body

Man and other animals live in an environment that has always contained poisonous chemicals. These occur most commonly mixed with foodstuffs and thus are likely to enter the body through the wall of the gut with the digested food. Such materials might include toxins present in the food materials (e.g., in fungi) or have arisen by spoilage of food (mycotoxins in grain or nuts). The body has various mechanisms to remove such materials (vomiting, diarrhea) or to process them in the liver to less harmful chemicals. Blood from the intestine passes first to the body's chemical factory, the liver, before distributing its load of nutrients to the rest of the body.

Natural poisons could also enter the body through the skin, although this is less likely because most toxins found in nature are solids that cannot pass through the skin unless in a solution. The skin is principally a defence against invasion by infectious microorganisms, as the intestine has to be also.

In the development of air-breathing animals, the lungs were the least likely organs to be challenged by toxins, because the air was free of poisonous vapours until the Industrial Revolution (apart from odd natural events such as the escape of carbon dioxide gas from Lake Nyos in Africa). Solid dusts, pollens, spores and similar substances have been challenges to the lungs, however.

Foreign matter can enter the body by various routes (Fig. 6.2). When we consider how chemical warfare agents might enter, we find a pattern the opposite of that for natural poisons. Entry through the gut is unlikely because a person will take care not to eat the agents or contaminated food. The skin presents an area of about 1.8 m^2 for entry (in an adult), and this is a vulnerable portal, but absorption through the skin is slow. A marginally fatal dose of nerve agent on the skin might cause death in thirty minutes, a large dose in one to two minutes. The lungs are designed for the rapid exchange of gases between the air and the blood (oxygen in and carbon dioxide out), so it is not surprising that chemical agents in vapour form enter most rapidly through these organs, allowing a lethal dose of nerve agent vapour to be taken up in one breath. These comments are presented in Table 6.1, which can be summarised by stating that the lungs have no natural

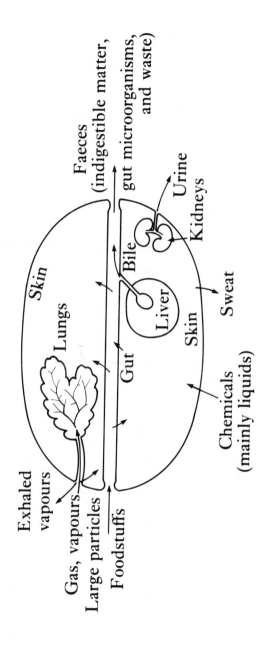

Figure 6.2. The main routes of entry into the body for foreign matter, and the paths for the expulsion of waste.

Table 6.1. *The effectiveness of mechanisms that protect the body against the entry of natural and synthetic toxic chemicals*

Entry portal	Protective mechanisms	Efficacy against natural poisons	Efficacy against warfare agents
Lungs	None for vapours, size filtration for particles	Not relevant – poisons are mainly solids	No protection against vapours, some against particles
Skin	Particles and solids cannot enter, liquids are slowed	Good	Slows entry of liquids, bars solids and particulates
Gut	Ejection (vomiting, etc.), selective absorption, detoxification in liver	Well developed and effective for most challenges	Not relevant

protection against gases or vapours, and the skin acts only as a delaying mechanism, rather than as a barrier, to liquid poisons. It is clear that the physical form of the warfare agent is important (Fig. 6.3).

Equipment for individual protection

It follows from the previous section that most attention must be paid to protection of the respiratory tract, followed by the skin. A range of specialised equipment is available for the protection of the lungs during industrial work, with filters to remove particulate matter (e.g., for work with asbestos or for floor sanding) or to remove a specific gas (hydrogen cyanide in plating shops, methyl bromide during fumigation). The military respirator has to protect against all possible battlefield challenges, particles, gases or vapours. The filter canister (Fig. 6.4) therefore incorporates both a particulate filter, now usually made of fluted glass fibre paper, and granulated charcoal to absorb gases and vapours. The charcoal has to be chemically treated to absorb low-molecular-weight gases, whose small molecules do not absorb strongly to surfaces. The filter canister is mounted on a facepiece of synthetic or natural rubber, which has eyepieces for vision and a valve system to allow inhalation through the canister and exhalation through a separate vent. Other fittings can include a voicemitter to render the wearer's voice intelligible to the outside listener, and a drinking tube that can be coupled to the water canteen. A typical military respirator is illustrated in Figure 6.5.

The early forms of field respirator were hurriedly developed in 1915 (Fig. 3.3) and by 1918 were becoming quite effective. Work on the protection of

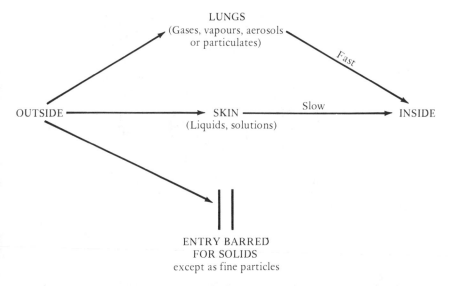

Figure 6.3. The routes of entry for chemicals into the body, related to the physical form of the particular chemical.

coal miners in the 1920s and 1930s culminated in the present basic designs, which were perfected for military use in World War II. It is commonly remarked that respirators are old technology or outdated, or it is said that World War II respirators would not protect against nerve agents. This is nonsense; the design is largely unchanged because the respirator is, and was, very effective. A general service field respirator from 1942 would protect against nerve gas as well as a modern one, for the molecules of nerve gas absorb well on charcoal. The only question about respirators is how well the filters absorb gases containing numbers of fluorine atoms in the molecule, for such gases do not absorb well. The compound PFIB (perfluorinated isobutylene) and similar chemicals are currently under investigation.

Experiments at the Materials Research Laboratory in Melbourne have shown that of a variety of types of military respirators, half or two-thirds do not fit well enough to protect the wearer in all conditions, even when the facepiece is correctly sized and carefully fitted. A minority are excellent in design. It is therefore essential to select a good respirator design; however, even when this is done, inexpert wearers will often not get a good fit to the face. Respiratory protection therefore depends both on good equipment well maintained and on thorough training of the wearer.

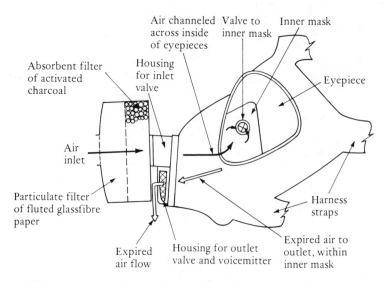

Figure 6.4. A drawing showing the functional parts of the military respirator or gas mask. Air is drawn in through the filter canister, across the eyepieces (to reduce fogging), then into the inner mask and the nose of the wearer. Expired air is discharged directly from the inner mask to the outside through the outlet valve. The canister is shown at the end of the mask (pig's snout conformation). It can also be fitted to the side of the facepiece, as in Figure 6.5.

The skin must be protected by clothing that resists penetration by liquids, vapours and particles. This may be a plastic sheet of suitable properties, a woven fabric supporting a plastic film, or an air-permeable fabric incorporating powdered charcoal to absorb vapour. The most common form of military protective suit consists of an inner, moderately thick layer of nonwoven fabric or polyurethane foam treated with a fluorocarbon repellent and coated with charcoal. Outside this is a thin layer of wear-resistant fabric, which is treated to repel water and (sometimes) liquid agent. Such suits protect well against likely field concentrations of chemical warfare agents. The protective ensemble is completed by the addition of gloves and overboots of resistant rubber.

This ensemble offers excellent protection but poses two problems: The wearer gets extremely hot if he does much physical activity, and movements are clumsy because of the hindrance of the garments and limitations to peripheral vision. The heat-stress problem is the major one, for clumsiness and the effects of limited sensory input can be reduced by familiarisation and training.

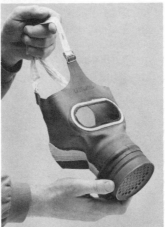

Figure 6.5. *Top:* The author wearing a modern military gas mask, the S 10 model produced by Avon Industrial Polymers of the United Kingdom. A novel feature is the recessing of the eyelens into the mask to enable the wearer to bring optical sights closer to his eyes. *Bottom:* The U.K. civilian respirator of World War II, similar to that shown in Figure 3.5. Photographs from the Materials Research Laboratory, Melbourne.

When working, the human body generates heat in exact proportion to the rate of working. Normally this heat escapes by convection, expiration of warm air and sweating. Once the body is enclosed, heat loss is reduced and therefore the body will heat up. The normal body temperature of around 37.4° C can rise to 38.5° C without causing much of a problem, but as it rises over 39° C there is a danger of a loss of temperature control altogether, followed by sudden collapse and death. The common military suits mentioned previously are made of air-permeable fabrics that allow some perspiration to evaporate and pass out and thus give a cooling effect. This is of reasonable value in temperate climates and in hotter climates in dry, moving air (say 37° C in a desert breeze). It is of no use when the air is stationary and saturated with water vapour, as under a jungle canopy. Figure 6.6 gives some illustrative results, showing how a person's body temperature can rise, and how long it can take to cool when activity ceases.

Mad dogs and Englishmen go out in the midday sun, rush around busily and drop from heat stroke. Queenslanders and well-trained soldiers in protective suits go out in the cool dawn, set a slow and deliberate work pace and get a modest amount of work done. It is often said that it is impossible to wear protective suits in the tropics, but this is not quite true. As with most things it is a matter of degree. If you compare a soldier in normal dress in a temperate climate with one in protective garb in the tropics, then the latter can do practically no work compared with the former. But consider a soldier wearing only shorts in 35° C heat, 90 percent relative humidity and no air movement. His work performance is not going to be much better than that of the soldier in the full rig; he has no mechanism of losing heat either. It is true that activity will be very much slowed when wearing protective equipment in the tropics; even an activity like patrolling slowly through jungle is virtually impossible, and similar activity in open, dry country is to be attempted with great care. You should also remember that strenuous activity in a cool climate can bring on heat stress when a person is fully encapsulated in protective garments.

Collective protection

There are two main reasons to have collective protection in a chemical battle area: to provide individuals with some relief from the burden of wearing individual protection, and to provide a protected environment for those persons who do not need to be mobile but who perform delicate tasks impossible in the individual protective ensemble. (These tasks include field surgery and care of casualties, and also the operation of a command and communication centre.)

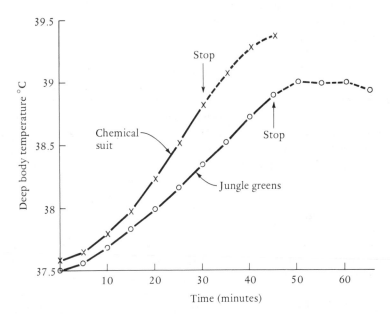

Figure 6.6. The graph shows the increase in deep body (rectal) temperature of volunteers doing moderate to heavy work. They were walking at a steady pace in a temperature of 30° C and 80 percent relative humidity, typical of a jungle climate. Each of the group of twelve persons carried 25 kg of military equipment, and wore either a chemical suit (similar to that in Figure 6.9) or their normal jungle greens of cotton denim. Note that in both types of garment they get dangerously hot, but at a quicker rate in the chemical suit. However, the greatest difference is in what happened when they stopped moving and sat down. In the chemical suit, the body temperature continued to rise sharply for at least fifteen minutes, and no cooling was seen. From data obtained at the Materials Research Laboratory, Melbourne.

The first requirement can be fulfilled with less stringent specifications than the second, as a useful degree of rest and relief can be obtained in thirty to sixty minutes, whereas a surgeon in battle may be working twenty-four to thirty-six hours at a stretch. The stringency refers to the residual level of warfare agent vapour in the "clean" atmosphere; the accumulated dose to the inhabitants of the protected structure depends on time, so shorter times can accept higher levels of agent.

The lower the permissible level of contaminant in the collective-protection shelter, the longer it takes to get people in, for a contaminated person will need a more thorough clean and decontamination if the shelter is a medical area with a stringent requirement. One great problem with collective protection is the question of entry; it may take fifteen minutes for a team of four to decontaminate a casualty to a degree that he could be allowed into a surgical unit.

There is a third use of collective protection whose utility really depends on how the military intends to fight in a chemical battlefield. This is a collectively protected vehicle such as an armoured personnel carrier (APC) or tank. The Russians have collective protection in the BMP series of APCs, and the Leopard tank of West German origin, now in service in Canada and Australia, also is fitted with collective protection. Such systems are good until the crew has to get out into the contaminated battlefield. It is believed that the Soviet philosophy for use of the BMPs was to drive them right through the contaminated zone as part of a blitzkrieg-type deep armoured thrust. A self-contained tank system is fine but goes against a cavalry tradition of fighting with the commander peering out of the open top of the tank.

All collective protection systems have the same functional elements: an electric blower, filter system, air conditioner (de luxe versions only), main compartment and entry–exit airlock (Fig. 6.7). The main compartment can be any existing structure that is rendered nearly airtight, but in field use it can be a fabric tentlike enclosure, either slung from a rigid frame or else having a double skin envelope that is inflated hard to give rigidity. The supply of filtered air from the blower has to exceed the leakage rate from the compartment and airlock, so that the interior is at slightly greater pressure than the outside, and consequently all leakage is outwards. The flow rate also has to be sufficient to provide fresh air for the occupants; the air must be changed at a rate dependent on the number of people in the compartment. Collective protection systems feature in the equipment plans of a number of armies, yet are not extensively deployed or used in training. Reasons for this may be that the equipment is bulky and costly and that the proposed plans for use are still matters of argument.

Early warning, detection and monitoring

It is desirable to have equipment that will warn of chemical attack (a chemical cloud upwind, or liquid droplets arriving on target), will enable the detection of toxic chemicals at dangerous levels at various locations and will make it possible to distinguish contaminated areas and equipment from clean areas after the attack. The latter function is of great practical importance, as it determines whether the gas mask can be taken off safely. The sensitivity of the equipment, therefore, has to be at a level that corresponds to safe unmasking. The technical requirements for all three functions are

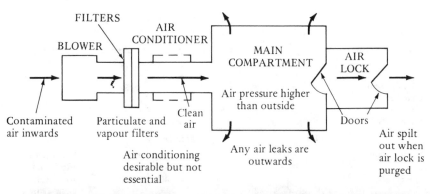

FILTERS

BLOWER

AIR CONDITIONER

MAIN COMPARTMENT

Air pressure higher than outside

AIR LOCK

Contaminated air inwards

Particulate and vapour filters

Clean air

Doors

Air conditioning desirable but not essential

Any air leaks are outwards

Air spilt out when air lock is purged

Figure 6.7. The diagram shows the layout of an idealised collective protection system. The photograph shows the interior of an actual collective protection installation, the Beaufort NBC (nuclear, biological, chemical) shelter. The framework supports fabric that resists penetration by chemical warfare agents, and the whole shelter is dug into the ground for protection against blast and fragments. Photo courtesy of Beaufort Air-Sea Equipment Ltd. Merseyside, United Kingdom.

Table 6.2. *The capabilities of detection equipment*
as a function of time after chemical attack

Equipment characteristics	Time after chemical attack		
	"Real" few seconds	Short seconds to few minutes	Long hours
Sensitivity	Good	Good	Excellent
Selectivity	Poor	Improving	Excellent
Requirement	Rapid warning of attack	Checking and monitoring after attack	Confirmation after event and forensic examination
Equipment	Alarms specifically for groups of agents. CAM in alarm role	Detector tubes and tickets, CAM	Laboratory-type equipment

Note: CAM = chemical agent monitor.

essentially the same: sensitivity, selectivity and rapid response. Sensitivity is not a problem with today's technology, nor in principle is speed of response. However, the compromises necessary to obtain selectivity mean that the other two factors are degraded. Table 6.2 shows that selectivity is dependent on time after the attack.

The simplest equipment is the detector tube, a narrow glass tube packed with an inert support medium, usually silica gel, in which chemical reagents that change colour on exposure to the agent vapour are impregnated. A standard volume of air is drawn through the tube by means of a hand pump, and the tube is then examined for a colour change. The length of any colour zone is a crude index of vapour concentration. These are single-use, expendable items, cheap, slow but useful. They have moderate sensitivity and selectivity.

The most sophisticated handheld device is the chemical agent monitor (CAM), which uses the principle of ion mobility spectrometry, a physical rather than a chemical method. Air is drawn through the sensor (Fig. 6.8) and the response of the intrument is shown on a liquid crystal display on the top; one bar indicates a very low concentration, eight bars is a maximal response. The monitor can be switched between nerve agent and mustard modes. It can also be tuned to respond to industrial chemicals and thus has an application in occupational hygiene.

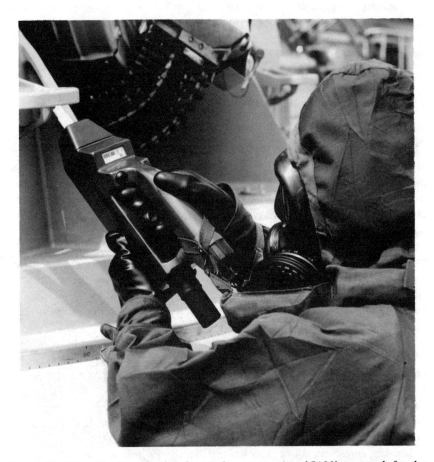

Figure 6.8. The sailor is using the chemical agent monitor (CAM) to search for the presence of chemical warfare agent vapour coming off possibly contaminated ship structures. CAM is the first real time monitor to have the required sensitivity for nerve agents and mustard. The principle of operation was the result of research by a team of British, American, Canadian and Australian scientists, and the instrument was commercially developed and produced by Graseby Ionics Ltd, Watford, United Kingdom, who supplied this photograph.

Many other forms of warning, detection and monitoring devices are available, including infrared devices that can detect clouds of chemical agent at a distance.

In general, early warning is difficult, and impossible if the target is attacked directly. Detection of the presence of agents is quite easy given the appropriate equipment, which may be cheap but slow (detector tubes) or expensive but near real time (CAM, approximately £4,000 per unit). Monitoring is best effected with the CAM, which samples air drawn directly from the suspect surface. Users of the CAM have to remember that it is a sensitive piece of scientific equipment, to be handled carefully and its output interpreted with care.

Decontamination

Chemical warfare agents that persist on surfaces because they are nonvolatile, stable chemicals can be removed by physical methods such as washing with water and detergent or with solvent liquids, or blowing hot air over the contaminated surface. Alternatively, the toxic material may be chemically altered by reactive chemicals to form less toxic products. The commonest process is to use solutions or mixtures of reactive chemicals (alkalis and bleaches) and also wash down, so that a hybrid of the two main methods results. The disadvantages of this are that the decontaminant liquids are corrosive to metals and injurious to plastics, and the method is labour-intensive as the solutions have to be brushed or sprayed over the vehicle or gun being decontaminated, unless steam or pumped sprays are available (Fig. 6.9).

Human skin obviously demands milder treatment, and the most widely recommended decontaminant is fullers' earth, the decomposed clay mineral formerly used for fulling or degreasing woollen cloth. As a powder this is reasonably effective at removing liquid from skin, provided that the action is taken within one minute of exposure to the toxic liquid. However, given the toxicity of VX or mustard, the complete protection of skin is infinitely preferable to attempts at removing contamination.

The big question about decontamination is: When is it to be done? It takes time and effort, which are not to be spared in battle, and when done, there is the possibility that all the clean equipment may be contaminated again in the next minute. The practical answer is to do only what is necessary to continue the operation, that is, remove gross contamination at points likely to be handled or contacted frequently. If equipment is withdrawn from battle to be repaired, then it may require complete decontamination.

Medical treatment in the field

The toxic effects of nerve-gas exposure are very rapid, and a chemical casualty may die in minutes or hours. The obvious symptoms of mustard poisoning are slow to develop, but the actual damage to the skin cells is

Figure 6.9. Steam lances are being used to decontaminate a military vehicle. The process is hard work, and can be more so if steam is not available with the result that decontaminant has to be brushed on by hand. Photograph from the Materials Research Laboratory, Melbourne.

irreversibly completed in minutes. This does not leave any time for medical intervention, and clearly the soldier on the battlefield needs treatment that can be self-administered or given by means of a buddy system.

The oldest form of therapy for nerve-agent poisoning is the injection of atropine, which does not reverse the poisoning, but suppresses its effects. In the mid 1950s the group of chemicals called oximes was developed, which can actually reverse poisoning by some, but not all, nerve agents. Atropine and oxime together are quite effective at saving people poisoned with moderate doses of GB or VX, but are less effective against GD and GA. The use of anticonvulsant drugs such as Valium also helps in treatment.

It was observed in the late 1940s that physostigmine, the active principle of the calabar bean, protects against nerve-agent poisoning, and Chinese field kits of the 1960s contained a similar drug, pyridostigmine. Later research has shown that carbamate esters, a class to which the two drugs belong, can protect a person against nerve agents if given before exposure, particularly when followed afterwards by therapy with oxime, atropine and

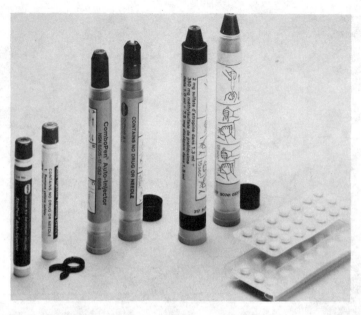

Figure 6.10. *Left:* Three types of autoinjector to administer nerve agent antidotes. The AtroPen is shown on the left, the ComboPen in the middle and the MultiPen at right. One of each pair is the training device, which strikes a blunt plastic button against the user's thigh, rather than a needle and injection fluid. The pack of tablets contains pyridostigmine doses for pretreatment against nerve agent poisoning. *Right:* This soldier is using the Combopen training aid to simulate self-injection into the thigh. Photographs courtesy of Duphar BV, Amsterdam, the Netherlands.

anticonvulsant. However, the protection is marginal, so that a person may survive, say, five lethal doses if treatment is given but will succumb to ten or more lethal doses.

In the field, treatment must be quick and foolproof; these requirements have been met by the design of autoinjectors (Fig. 6.10), which rapidly inject drug solutions into the muscles (thigh, buttocks) of the poisoned person. Atropine and Valium have side effects of their own, so that if taken in error as a result of a false alarm, the soldier can incapacitate himself or his mate. The rules for self-injection are therefore difficult to formulate; to my mind it is impracticable and best done by a third person on the basis of some objective signs.

The medical staff at the forward aid post will receive chemical casualties, conventional casualties and the doubly unfortunate who have both prob-

Figure 6.10 (*continued*).

lems. An overloaded medical chain will then have to pass to the rear those
with a chance of survival. The most difficult requirement would be to provide artificial respiration to numbers of nerve-agent casualties, who may be
entirely dependent on this aid.

There is no specific treatment for mustard injury; the burns on the skin
must be kept sterile and treated in the same way as for thermal burns. Persons who have inhaled mustard vapour will die slowly.

Medical treatment provided quickly can save persons exposed to nerve
gases, but those who have received large doses will die quickly. There is no
magic cure; the effect of treatment is to save those who have been marginally exposed. Each advance in treatment has been hailed as a miracle that
will render protective clothing unnecessary and nerve agents obsolete.
Would that it were so, but the reality is that each improvement is a small
step to limited success.

Banning chemical weapons

Table 6.3. *The major problems in the defensive techniques of chemical warfare*

Technique	Problems encountered
Individual protection	Cumbersome to wear and movements clumsy, work induces heat stress, psychological effect of feeling of isolation.
Collective protection	Equipment is rather bulky to carry in field, decontamination of persons on entry is expensive in time and labour
Detection	Early warning is difficult, selectivity and response time are problems, equipment needs to be operated by well-trained persons.
Decontamination	Process is expensive in time and labour, decontaminants are often damaging to equipment, skin decontamination must be done very soon after contamination
Field medical treatment	Is marginally useful in reducing severity of intoxication, no specific treatment for mustard exposure, self-aid or buddy-aid has problems in the determination of when aid is to be given

Limitations and achievements of chemical defence

Some of you will have skipped the foregoing technical detail, wondering what relevance it has to the main argument of chemical disarmament. It is necessary, however, to understand something of the technicalities to gain an understanding of the attractiveness or otherwise of chemical warfare to an aggressor. There is no doubt that the defendants in a chemical war do suffer from disadvantages, not present in conventional war. Table 6.3 lists some of the main problems with the defensive techniques I have described. Probably the most emotive is the wearing of protective clothing by the individual, but this is also one that has no magic technical solution; it is largely psychological and physiological, and can be side-stepped to some extent by good training.

We come back to my statement that defence against chemicals can be effective, given good equipment and extensive training in its use. Few nations meet this requirement in their military forces, and none in terms of the civilian population. It is therefore reasonable to believe that an evil-minded aggressor could see chemical weapons as useful. This aggressor would need to invest heavily himself in defensive equipment, if only to protect against the aftereffects of his own weapons.

7

◁═══════════════════════════════════════▷

Discussions on matters
of particular interest

A number of topics deserve more attention than they receive in my more systematic discussions in the other chapters. Here I present short essays on topics of particular import, or on which I feel there is misapprehension by the public. These discussions are intended to complement or reinforce observations that appear elsewhere in the text.

Protective clothing: an impossible burden?

Sometime ago I was involved in a trial to ascertain how long soldiers could perform warlike duties in the Queensland spring, when wearing fully protective clothing. The regional commander landed in his helicopter, walked around for twenty minutes, said "My men will never wear this" and got back into his helicopter to depart to more conventional duties. We were left with the unspoken questions, such as "What were his men going to do if they had to face an enemy with chemical weapons?" (This was before the use of chemicals by Iraq against Iran.) Or "Can his men wear this and fight a war?"

The latter question in various forms is an endless topic for argument, the reason being that in recent times no one has tried it. By this I mean that the training in protective clothing that a number of armies have undertaken has not been of sufficient duration or realistic enough to test the matter. Although I have mentioned in Chapter 6 the problems of heat stress and hindrance afforded by protective clothing, it seems worthwhile reexamining this central question here.

First, there is the question of the reality of tests and trials. To some extent, the use of riot-control agents will make the trainees keep their respirators on, but they know that they face no real danger. There is nothing like the prospect of death to concentrate the mind wonderfully (thank you, Dr. Johnson) and that fear from warfare chemicals has not occurred much since World War I. I am sure that in the real situation, soldiers will find they can wear protection much longer and also perform military duties, at a reduced rate.

Be careful how you interpret that statement. One limiting factor to the time that protection can be worn, which is the buildup of heat in the body, is not influenced at all by psychological factors, fear of death, military morale or the last refuge of a scoundrel (again thanks, Dr. J.). It is a simple physical balance; if the production of heat within the body exceeds the loss of heat to the outside, then the body will get hotter, to the point that the person collapses of heat stress. What I mean by an improved performance in the real situation is that the soldiers will pay heed to what they are told, profit very rapidly by the mistakes of their fellows, and learn to match their physical exertion with the heat stress they experience.

The simple equation of greater heat stress with higher environmental temperature when one is wearing protective clothing is not true or, more correctly, it is the lesser truth. The dominant truth is that heat stress increases with the work rate so that heat stress is possible in the Antarctic if a person cannot discard outer clothing and is working hard enough. The textbook graphs of some expression of heat buildup in the body rising with external temperature are based on very wide temperature ranges. My advice to a military commander is that in any one theatre of operations, diurnal temperature variations will have little effect on stressing the protected soldier. By carefully regulating the rate at which soldiers work, the commander has control of potential heat stress.

The foregoing is true only of the protected, fully encapsulated person, because the body is then fully surrounded by a trapped envelope of humid air. The nature of the fabric that encloses it is of little consequence, for this air layer is the insulant. Also the main route of loss of heat when other mechanisms are inadequate is by the evaporation of sweat. This is almost completely inhibited by protective clothing. Take off the clothing, or some of it, and the situation entirely changes (Fig. 7.1).

In summary, the influence of climate is broad, that of rate of working is decisive. The answer to the original question is that protected troops can fight a war in hot climates, at a much slower pace. In climatic extremes this pace may be zero, but in these conditions, when troops are unprotected, the pace would also be slow.

One other point: Lost water must be replaced. Perspiration rates can be at 2–3 litres per hour for moderate work in hot climates.

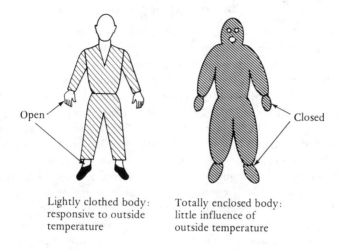

Lightly clothed body: Totally enclosed body:
responsive to outside   little influence of
temperature   outside temperature

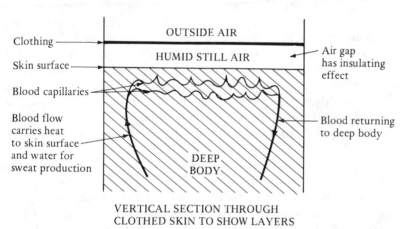

VERTICAL SECTION THROUGH
CLOTHED SKIN TO SHOW LAYERS

Figure 7.1. The influence of clothing on the loss of heat from the body. Normal light clothing (upper left) results in the heat stress to the wearer varying both with external temperature and the rate at which the wearer works. When the person is totally enclosed, the stress is mainly dependent on the rate of working, and is much greater than for the lightly clothed person. The main problem (bottom) is the layer of trapped humid air between the skin and the clothing, which prevents the evaporation of sweat and also the transfer of heat to the outside.

The civilian population

American and British peoples have an unreal attitude to war, which is the result of their history. War is something to which one sends one's fighting men and women, it is not something that comes to you. Thus war consists of assembling an expeditionary force and sending it overseas. Implicit in this concept is the belief that the civilian population is relatively safe. This attitude stems from the absence of major armed conflicts in the United States since 1865 or in England since 1660. Yes, I have myself pointed out the perceived susceptibility of the English population to air attack in the 1930s (Chapter 3) and the violent demonstrations of attack on civil targets given by the blitz on Coventry and London. Nevertheless, I do not think that American or British people have the same awareness of war that the French or Germans, for example, have got from their history. Gustavus Adolphus, Marlborough, Napoleon, Wellington, Frederick the Great, Kaiser Wilhelm and Hitler have all sent their armies through Europe, incurring as much, or more, damage to civilians than to the opposing armies.

The point I wish to make is that Anglo-Saxons tend to consider the effects of weapons on their armed forces, and only secondarily on the civilians. Yet it seems to me that the modern ability to strike far in the rear means that the effective way to win is to destroy all supply lines, assembly points and storage depots. It does not matter much what the enemy's front-line troops do after that. They can wander around a bit until their supplies are gone, but sooner or later they must accept the inevitable and surrender. This was obviously the tactic of the Multi-National Force in the Gulf in February–March 1991.

If conventional weaponry, with at least some potential for direction to a target, results in great "collateral" damage, what can we expect from chemical weapons, which are area-coverage devices and not capable of precise target selection? The answer must be great loss of civilian lives. No matter how good the protection for soldiers can be (Chapter 6), we cannot expect civilians to be well protected even if equipment is issued on a massive scale as in the United Kingdom in the 1930s. The effects of mustard gas might be markedly reduced, but to survive a nerve-gas attack would require a high degree of training and discipline.

I have seen a number of articles with diagrams showing the areas in which casualties would occur after the use of chemical weapons in Central Europe. The effects might well extend 50 or 100 km downwind of the release point, but they are entirely dependent on weather conditions. The accidental gassing of large numbers of people at Bhopal in 1984 was such a major disaster because of meteorological conditions at the time – at night in still air. A release of the methyl isocyanate in daytime in a turbulent updraft of air would have been much less costly of lives.

The endeavours to obtain a Chemical Weapons Convention are thus driven by concerns for the civil population; the military problem is secondary.

Hydrogen cyanide

This is regarded by many people as the best example of a poisonous gas, and indeed it is poisonous. Deaths from the inhalation of the vapour occur regularly due to leaks or accidental releases in industry. In the context of war gases, however, it is not a useful example. It is about fifty times less toxic than nerve-gas vapour, and the vapour is slightly less dense than air. The latter property means that hydrogen cyanide will disperse relatively quickly. It was used in World War I, but the general conclusion was that it had no military usefulness, compared with phosgene, for example (see Table 3.1). Cyanide continued to be of military interest until nerve gases were discovered, at which point it became, by comparison, ineffective.

Interest in hydrogen cyanide has been awakened by Aubin Heyndrickx, formerly a professor of Ghent University in Belgium. He has claimed to have found forms of cyanide in munition cases in Angola and has used this to support claims of chemical warfare there. I find his observations to be difficult to interpret however, and his statements about forms of cyanide are scientifically inexact. Further, the professor's credibility has suffered in a wider sphere, for his suggested procedures for treatment of chemical warfare victims, which were published in *Lancet*,[1] drew a number of critical comments from different experts in a variety of countries.

Hydrogen cyanide is not a prime choice for a chemical warfare agent. Nevertheless, because it is available in quantity from the chemical industry, it might appeal to a country that wished to acquire a chemical weapon in a hurry.

Residues

What would be left behind after a chemical attack? This question is important in relationship to, first, the remaining hazard to people and, second, the ability to determine if indeed an attack had occurred some time previously. The chemical can be dispersed in three main forms (Fig. 7.2), namely vapour, fine liquid drops and large thickened blobs. Vapour will disperse rapidly and cannot be seen (except for high concentrations of phosgene, chlorine and a few other gases) so that its presence can only be determined

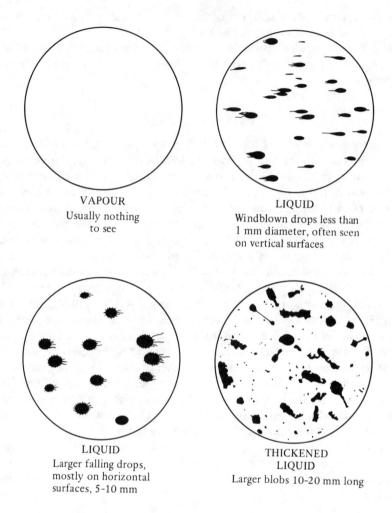

VAPOUR
Usually nothing
to see

LIQUID
Windblown drops less than
1 mm diameter, often seen
on vertical surfaces

LIQUID
Larger falling drops,
mostly on horizontal
surfaces, 5-10 mm

THICKENED
LIQUID
Larger blobs 10-20 mm long

Figure 7.2. The appearance of chemical warfare agents immediately after an attack. Not to scale. Fine liquid drops can be likened to spray from an atomiser, falling drops to rain and thickened drops to honey dripping from a spoon.

by detection equipment. It neither presents a continuing hazard to people nor leaves a permanent trace.

Liquid droplets will persist for a period dependent on the volatility of the particular liquid, but they will be small and colourless, very hard to see. A liquid like VX will persist for a very long time and thus present a hazard to persons for perhaps days or weeks, as will mustard. The vapour hazard from these persistent liquids will be a problem for persons who stay in the area, but very short exposures (minutes) should not be fatal. On the other hand, any transfer of the liquid to the skin will be very dangerous, particularly if it is VX, and the likelihood of this happening will of course depend on whether the liquid has soaked into the ground, vegetation or surrounding objects, or whether it is still present on the surfaces. Moderately volatile liquids will present vapour and liquid hazards for shorter periods of hours to one or several days. There should be no difficulty in determining if a liquid chemical attack occurred on an area in the past even if the vapour and the surface liquid have gone. Modern analytical techniques can find traces of the chemicals that have soaked into permeable surfaces and thus to an extent become trapped. The materials to be collected as good samples include artificial materials such as rubber, plastics, concrete or textiles, and porous natural objects such as bark, dead wood, dry leaves and soil.

The sticky blobs of thickened liquid should be relatively easy to see, even after much of the liquid has evaporated. The hazard is the same as for persistent liquid drops, bearing in mind there will be fewer, but larger, thickened drops. A person sampling the area would aim to collect the actual drop residue, rather than assuming an average contamination of all surfaces. Given a good sample, the analyst will have no problem identifying the warfare agent that was used.

We must remember that besides the actual chemical, there should be spent munition cases contaminated with residues of the chemical fill. Only in the case of aerial spraying will these be absent. Munition cases present a trap for the analyst, however, because residues from other chemicals apart from the toxic one will be present. The shell or rocket has to be propelled to the target, so that propellant residues will be left on the casing. Further, if there is an explosive charge to burst open the munition casing and disperse the contents, then there will be explosive residues. These propellant and explosive residues have been subject to thermal and chemical decomposition, so that they will contain a vast number of chemical compounds in trace amounts, some of which could be mistaken for the residues of toxic chemicals. It is also possible to mistake a conventional munition for a chemical one.

The question of residues is important to a Chemical Weapons Convention, as that will need mechanisms to detect violations. This is also discussed in Chapter 9.

Volatility and thickening

As discussed in Chapter 6, the volatility of a warfare agent is an important characteristic, as it determines how the chemical can enter the body. Any specific chemical has a given volatility—that is, the ease with which it turns to vapour—which is dependent on the temperature, but cannot be altered by human intervention. This is not a great problem, provided you have a choice of nerve agents, for you can choose from the nearly involatile VX through GF and GD to GB, which is as volatile as water.

Although one cannot alter the volatility of any one chemical, one can alter the rate at which it evaporates in a specific situation, and this can be done by thickening. A toxic liquid in a shell or bomb case can be dispersed by being blown apart by an explosive charge, or by the airstream around the travelling projectile. These effects tend to reduce the bulk liquid to fine droplets, which may be too fine to fall on the intended target. This effect is overcome (Chapter 5) by thickening, which produces larger blobs of liquid that fall faster. Moreover, they evaporate more slowly as the surface area of a given volume of liquid is smaller if broken into larger drops. Thus there is some indirect control over the rate of volatilisation. Moderately volatile liquids that are thickened rapidly form skins over the droplets, which also string out like spiders' webs, so that what hits the ground resembles blobs and strings of sticky treacle. The skin once formed also slows evaporation.

From the point of view of disarmament, the presence of the technology and means to thicken toxic liquids is indicative of a military end use.

Toxins

The difficulty with these is in defining them. In fact there are two definitional problems, the first being that many people are now using the term "toxin" to mean any toxic chemical of whatever origin. This has become popular recently among commentators on environmental topics, who refer to synthetic pollutants such as PCBs (polychlorinated biphenyls) or PAHs (polyaromatic hydrocarbons) as toxins. The second difficulty is that the term was used originally to signify a toxic product from a natural source— for example, tetrodotoxin from the puffer fish and blue-ringed octopus, or the various components of the venom of snakes. Now that many of the simpler toxins can be made synthetically, this distinction between natural and artificial sources is blurred. In fact the blurring commenced long ago, in 1828 when Wohler synthesised urea and the world then realised that there was no fundamental distinction between components of living systems and those made artificially. The very complex protein toxins have not yet been synthesised, but there is no reason why they should not. The largest toxin yet synthesised is palytoxin carboxylic acid, found naturally in a Hawaiian coral. This molecule has a core chain of over 120 carbon atoms bearing oxygen and hydrogen atoms and one nitrogen atom.

The original meaning of "toxin" is now irretrievably debased, and it would be best not to use it in situations where the meaning could be ambiguous.

Yellow Rain

It is remarkable how we do not notice what is familiar, until some outside event makes us look again. My car is kept parked in the driveway of our house, next to a rather poorly kept garden. Quite often, as I prepare to drive to work, I see one or more yellowish spots on the windscreen, about 5–7 mm in diameter. These are firmly cemented on, as it takes two or three passes of the wipers to get them off, after they have been wetted. The droppings of flower-feeding insects, they consist of digested pollen grains and other refuse. I have seen a hover fly deposit such a spot on a vine leaf. The yellow droplet formed at the rear of the abdomen as the insect hovered 3 cm above the leaf then fell off onto the leaf and spread out on the surface. It soon dried to a yellow spot.

Some years ago, however, I was presented with specimens of similar spots on leaves and pebbles and failed to recognize them. I must have seen the spots on the windscreen by that time but had never speculated as to what they were. The spots on the leaves and pebbles were also pollen grains in various degrees of digestion. Yellow Rain is the material that Hmong tribespeople in Laos reported in the late 1970s and early 1980s as being used to attack them chemically. However, the only connection between the yellow spots and chemical attacks is the association with yellow. When samples of Yellow Rain were requested, the insect droppings were produced as the only yellow material that could be found. No toxic material could be identified in any of the yellow-spot or yellow-powder samples by the British government or U.S. army laboratories, despite claims that mycotoxins were found in a few early samples. The result is that we have three entirely independent observations:

1. Hmong tribesmen were harassed into leaving Laos by various methods, which may have included the use of crop-destroying sprays, riot-control agents or other chemicals.
2. The droppings of flower-feeding insects, in particular honeybees, can readily be found in Laos and northern Thailand.
3. The fungi that produce trichothecene mycotoxins are widely distributed, and leave traces of their toxins on any material on which the fungi can grow, including human foodstuffs. People who subsequently eat the food may have traces of the toxins in their bloodstream.

There is no indication of any connection between these three observations.

These observations are simply a caution to think carefully about allegations of chemical warfare before drawing conclusions. The false trail of the yellow spots so confused the situation that we do not know now what, if anything, really happened in Laos.

Agent Orange

Most people I speak to confuse Yellow Rain with Agent Orange, but have a much better knowledge of the latter. Orange was the colour code for the mixture of 2,4–D and 2,4,5–T herbicides, which formed one of the several defoliants that were sprayed over Vietnam to reduce the cover afforded to the Viet Cong by the forest trees. You are all familiar with the still-unresolved dispute as to whether veterans of the war suffered health defects due to exposure to the herbicide. However, that dispute is not central to the chemical disarmament question, but Agent Orange does introduce another key problem that needs resolution. Agent Orange was a chemical used for military purposes in war. Is it a chemical warfare agent? This is a problem of definition I have considered elsewhere, principally in Chapter 9, but it is worth thinking about here.

Blue X, happy gases and psychoactive drugs

There were stories from Afghanistan in the 1980s about the use of a gas that made the Mujaheddin lie down and sleep tranquilly until they awoke hours later in Soviet custody. Because of reports of a blue colour, this became known as Blue X. No hard evidence was ever produced for the existence of this gas, so there is a strong suspicion that some unusual event has been grossly exaggerated for propaganda purposes. Independent of Blue X is a belief in gases that make people euphoric to the point where they may become careless enough to cast off their protective clothing, thus exposing themselves to more lethal chemicals. I do not know of any such "happy gas" that has been shown to be militarily effective.

General interest in LSD and other drugs that had powerful effects on the brain, that could generate hallucinations or otherwise disconnect the victim from reality, stimulated the military to examine the battlefield potential of such drugs. This culminated in much research in the 1960s, but it was generally agreed that the great variation in individual response to such drugs meant that the results of their use were totally unpredictable and therefore of no practical use. The ideal would be a drug that put the victim quietly to sleep or in a trance state, from which he could be aroused by administration of a second drug. Nothing like this was achieved. The United States did put a psychoactive drug, BZ, in its chemical armoury, but this was destroyed in the 1980s. BZ effects were unpredictable; as for LSD the effects depended very much on the existing mood of the individual and on the surroundings. Crude analogies can be made with alcohol, which may act as a soporific but in the right atmosphere may also act as a stimulant to excessively boisterous behaviour.

These compounds have no current military significance but do have an impact on a Chemical Weapons Convention if such drugs are possible developments. This is another occasion where we have to include in our definition of chemical warfare agents a chemical that has no long-lasting effect on the victim (Questionable for LSD at least) but might have a military use. The situation is analogous to that of riot-control agents. (See Chapter 5.) Fortunately, in practical terms the control of psychoactive drugs would not be too difficult, for any conceivable peaceful use would require very much smaller amounts than would be required to spread over a battlefield. A nation holding stocks of hundreds of kilograms of such a chemical would have to be challenged as to the end use. Remember that we are talking about drugs for which the human dose (in controlled conditions) is a few micrograms.

Reference

1. Heyndrickx, A., and B. Heyndrickx. (1990). "Management of War Gas Injuries." *Lancet* (17 Nov. 1990): 1248–9. The comments by a variety of authors on this article appear in *Lancet* (12 Jan. 1991): 121–2.

8

The chemical industry and trade in chemicals

In times of change, some changes are dramatically apparent and capture our attention, whereas others, perhaps of greater significance, arrive and establish themselves unseen and unremarked. In my lifetime, the technical developments most often commented on are nuclear bombs and power generation, space travel and computers, all the result of advances in physical science and technology. Yet a parallel and all-pervading growth in the use of synthetic chemicals has also occurred. Some aspects, such as environmental pollution by chemicals, are much in the public consciousness, yet the central position of chemicals in the economy of late twentieth-century life is not well appreciated. The easiest way of showing how they pervade modern society is to look at the recent past in a series of freeze-frames or snapshots.

Frame 1890: The practical chemistry of metal refining and ceramic production is understood, as is the modification of natural products, such as the formation of soap from fats or the tanning of hide to form leather. The production of basic chemicals for these processes is a big industry, but synthetic chemicals are represented mainly by the dyestuffs, first synthesized in Britain (mauve was made by W. H. Perkins in 1856) but now dominated by a German dyestuffs industry, which is using the expertise acquired in this field to advance into other areas. Pharmaceutical drugs are all natural products, purified and standardized by the methods of the various pharmacopoeias. Materials of construction, fabrics, fibres and so forth are metals and ceramics or plant or animal products. Groceries and produce are sold in bulk, the only containers being glass bottles, tins and paper or fibre bags.

Frame 1930: There is a developing industry making the first synthetic polymers, such as bakelite, and urea-formaldehyde and alkyd resins. The application of synthetic chemistry to pharmaceutical development has produced Salvarsan for the treatment of syphilis and aspirin for the treatment of fever and inflammation. However, there is still great dependence on natural products.

Frame 1940: New polymeric materials are being developed, which can be used for structures, sheets and fibres, such as polythene, polystyrene, PVC, synthetic rubbers and nylon (marketed in 1941). The pharmaceutical industry is increasingly examining the modification of existing natural drugs, including the newly discovered antibiotics. Synthetic chemicals are being investigated for use in controlling agricultural pests, and in Germany the new organophosphorus insecticides are being developed into war gases. The year 1940 is truly on the threshold of the New Chemical Age,[1] which is receiving an impetus from the needs of the war now beginning.

Frame 1950: Plastics are taking over many of the roles formerly filled by natural products, and the range has increased to provide structural materials, fibres, wrapping films, sheets, paints, varnishes and so on. There is a dramatic development in the use of synthetic chemicals for pest control, and there is hope that famines can be prevented and diseases eliminated by their use (in the latter case by the control of the insect vectors of malaria, yellow fever, etc.). The introduction of plastic films and bags is causing a revolution in the packaging industry; suppliers are packing grocery items in individual containers with brand names, rather than forwarding bulk supplies to the retailer. The grocer no longer weighs sugar into a paper cone for the customer. The source of these new materials is largely coal or modified natural products (e.g., rayon), but oil is starting to be used as the base material.

Frame 1990: Synthetic chemicals now totally dominate those material requirements not filled by metals, ceramics and the declining natural products (wood, animal and vegetable fibre). My room is carpeted with wool/nylon fibre backed with a synthetic latex. The wooden desk is surfaced with a polyurethane varnish on which I have a vinyl mock-leather writing pad. The walls are papered with a paper that has an acrylic surface to reduce soiling. The keyboard and case of the word processor are made of a rigid plastic and sit on a stand of chipboard coated with a plastic laminate bonded by a contact adhesive. The electricity is brought to the light through copper conductors insulated with PVC. My car is a combination of metal and synthetic polymers. If I become sick, it is quite likely the drug I am treated with is a synthetic one, bearing little resemblance to a natural product. I would receive an injection from a disposable plastic syringe, or an intravenous drip through plastic tubing, perhaps containing a synthetic polymer as a component of a plasma expander. Every item in the supermarket is already individually wrapped, in polythene or other polymers, sometimes combined with paper or card. In other words, synthetic chemicals dominate our material needs. And where do they come from? From oil, via an enormous and complex petrochemicals industry. Since 1950, the growth of this industry is perhaps the major factor in the world's manufacturing economy. As events in the 1990s are proving, we are dependent on material taken from the ground in Kuwait or Saudi Arabia—not just to meet the immediate need for motor fuel but for all our material needs.

Are we getting far away from the topic of chemical disarmament? No, for a real appreciation of the modern chemical industry is necessary if we are to devise methods that ensure it cannot be turned towards the manufacture of chemicals for war. Frame 1990 should make you realize just how much we rely on synthetic chemicals in normal life, and you will appreciate that this all-pervading production system needs a very complex trading and supply system to get the products to market.

The major chemical producers are the United States, Japan, Germany, the United Kingdom, France, Italy and the Netherlands. Other countries have large producers of specialty products, for example Switzerland has Hoffman—La Roche (pharmaceuticals) and Ciba-Geigy (pesticides, dyestuffs, Araldite). The petrochemical industry is of course dependent on a supply of crude oil, which accounts for the strength of the U.S. industry, for historically it has had a domestic supply of oil and natural gas. This crude is turned into primary products such as bulk polymers and solvents, which can be sold off into the international market, or turned into the final product on site. There is an increasing trend for the oil-producing countries to do the first synthetic steps, for then they receive the advantage of added value with lower freight costs. Thus Iran, Iraq and Saudi Arabia are building up petrochemical plants to produce, for example, bulk polythene from the crude oil. Chemicals quite often have a limited number of suppliers, one or two in each country, so that there is a trade in chemical intermediates, which are further processed to produce a range of final products. Thus a factory producing an insecticide or herbicide may just do one or two final steps in the synthetic process, having bought the intermediate chemicals on the market. It will then produce the final formulation—add wetting agents, dispersants or whatever to the active ingredient, pack the product and sell to the wholesaler. Very complex final products, such as pharmaceuticals, are supported by a trade in relatively small quantities of quite sophisticated intermediates, and in this case there may well be just a few suppliers worldwide. It is therefore quite clear that the chemical market is a world market, requiring the shipment of chemicals from country to country depending on the complex patterns of production and marketing.

It is also evident that the administrative and financial control of this trade is international in character. Indeed, a chemical company in one country may control the movement of a chemical shipment in the other hemisphere, without the material ever being near the home country. Further, for various reasons, which include minimising taxation and evading national laws regulating cartels, it is often desirable for a chemical company to operate through offshore subsidiaries and to maintain marketing branches ostensibly independent of the parent company. A national government can therefore control activities within its national borders, and monitor the movement of chemicals inwards and outwards, but it cannot control the final destination of exports, nor by itself can it see the whole picture of world trade. An international body to monitor the movement of chemicals will be necessary if and when there is a Chemical Weapons Convention, a body with power to control not just weapon chemicals but the precursor chemicals from which they can be made.

What is also necessary is the cooperation of industry itself. The major industry bodies have shown themselves to be quite willing to assist national governments and the United Nations in devising methods to control the movement of chemicals towards nations or firms that might be involved in chemical weapons manufacture. The Government–Industry Conference against Chemical Weapons, held in Canberra in September 1989, was a step towards achieving mutual understanding between industry and government. Industry requires assurances that unfair or arbitrary restrictions on trade will not occur, that the need for commercial confidence is respected and that it will not suffer a bureaucratic imposition of paperwork and pointless questions. Government needs to put in place a control system that really works and is seen by industry to be acceptable. The chemical industry does at present have to supply much information to government. At the factory level, they are subject to various regulations about the safety of workers and about environmental pollution, as a consequence of which they may have to report to government on the nature of their activities, their products, waste products and so forth. They are also subject to inspection by government bodies. On a national scale, the company has to make returns to government in relation to the tax of profits, payroll tax, other excise or sales tax liabilities and similar matters. Governments therefore already know a good deal about their chemical industries, and may not need to ask for much more if they can usefully coordinate the information they already have.

Large companies and trade associations see the negative public image that would arise from any complicity in chemical weapons manufacture, but smaller companies may see a profit to be made. The main danger thus comes from a chemical trading company set up specifically to move chemicals from place to place until the origin and destination are both indeterminate, except to the operators. Such a firm can operate virtually independently of any national authority, so that control can only be exerted by the vigilance of the original suppliers, reporting to an international authority, and of the customs officers at points of export and import. The latter may not have the power to prevent movement of the chemicals, nor a reason to do so, but reports on suspicious transhipments can be assembled into a picture of illicit activity or of innocent trade.

In my opinion, therefore, it is the complexity and international nature of the chemical industry and trade that need to be understood if we are to make sensible moves to prevent chemical arms manufacture.

The desirability of monitoring the international trade in weapon chemicals and their precursors was recognized in the mid-1980s, largely as a result of the continued manufacture and use of such weapons by Iraq. An informal meeting of countries in the Western bloc with chemical industries of significance was chaired by Australia, and such meetings have continued since. The Australia Group has no formal structure, but consists of diplomats, their scientific advisers and members of the intelligence community. They consider reports on the movement of chemicals from government and commercial sources, with the objective of identifying trade patterns that might show a country is endeavouring to make weapons. The group has also developed a list of chemicals that are of interest, so that member countries know which are the chemicals whose movement should be notified to the other group members. Some countries have enacted legislation to require formal controls on certain chemicals. The Australia Group therefore depends on governments and industry to inform it voluntarily of suspicious trade; to an increasing degree this control is backed by legal sanctions. Australia introduced national export controls on eight chemicals in 1984; the number increased to thirty chemical weapons precursors in 1987, and in 1991 a standard list of fifty chemicals was recommended to participating countries. Controls on the movement around the world of chemicals that have military implications can work only if officials are vigilant and merchants are both informed and sympathetic. An example of what can happen is furnished by the export from the United States of thiodiglycol, a direct precursor of mustard gas, which was done in breach of U.S. government regulations and resulted in successful prosecution by the U.S. Customs Service. The firm of Alcolac International was approached in 1987 by two sets of persons representing the two combatants in the Iran–Iraq War, each seeking to buy thiodiglycol. The Iranian approach was through a West German firm known as Colimex, whereas the Iraqis used a New York firm, Nukraft, as a cover for a Dutch person working for them. The Iranians received three shipments of thiodiglycol, one through Greece, and two through Singapore, Hong Kong and Pakistan successively. Or they would have got three shipments if U.S. Customs agents had not substituted water for the 120 tons of the third shipment as it left Norfolk, Virginia. The Iraqis got four shipments from Alcolac before that company became more cautious, the route of supply being via Belgium and Jordan to Iraq.

This supply of war material contrary to U.S. bans happened because the supplying company, Alcolac, turned a blind eye towards very suspicious requests and also helped to produce misleading shipping documents that enabled the intermediaries to hide the country of ultimate destination. The company sustained a fine of nearly U.S. $1 million. The company could hardly have not noticed the unusual nature of the requests from the outset, as the total quantity of thiodiglycol supplied to the two channels in 1987–8 was about 700 tons, against a usual yearly sale of around 900 tons. Such an increase in demand indicates an abnormal factor has entered the market, and this gives the regulators a further method of controlling chemical movements. Any unusual patterns of trade in chemicals of military interest must be investigated.

A system of export controls cannot stop a country from making weapon chemicals, but it can make things more difficult and delay production. This seems to have been the effect on Iraq; supply became more difficult, particularly as some European firms have been prosecuted by their national governments for supplying thiodiglycol, other chemicals and chemical processing equipment. Iraq had the choice of making all the chemicals from the simplest starting materials, or buying more remote precursors from overseas before the 1991 ceasefire stopped weapons manufacture. The problem for the scientific advisers is to guess in which direction the suspect country will move its chemical production. Will it start to produce a new chemical agent? Has it discovered a novel synthetic route to produce nerve agent? How good are its chemists, and what technical information does it possess? Of course, the two countries of most interest have been Iraq and Libya; the other that may well be of concern is Iran, but unequivocal proof of a chemical weapon capacity is lacking, although strongly suspected. Still others, such as Syria, may well have weapons but keep a very low profile.

The monitoring of industry activities in detail, by inspection of factories and store places, is another activity that has to be planned carefully, and this is discussed in Chapter 9.

If the reader wishes to obtain more detail on the modern chemical industry, the best sources are the two books edited by Heaton.[2,3]

References

1. Crone, H. D. (1986). *Chemicals and Society: A Guide to the New Chemical Age.* Cambridge: Cambridge University Press.
2. Heaton, C. A. (ed.) (1984). *An Introduction to Industrial Chemistry.* Glasgow: Leonard Hill.
3. Heaton, C. A. (ed.) (1986). *The Chemical Industry.* Glasgow: Blackie.

9

Technical problems in the implementation of a Chemical Weapons Convention

One of the reasons for attempting to ban chemical warfare is the feeling that it is a peculiarly terrible way of prosecuting war. Mustard gas victims die slowly due to asphyxiation as their lungs are destroyed, at the same time suffering extreme burns on their skin. Phosgene victims also die slowly as their lungs fill with fluid and again asphyxiate them. The victims of nerve-gas poisoning may die very quickly after inhaling a large dose, or may hover between life and death after a marginal dose. Certainly this is terrible, but is it more so than the injuries caused by "conventional" (and, by implication, acceptable) means of war? A medical text about the treatment of mustard gas injuries in World War I[1] has a plate showing the head of a man who, although still alive, has had all of his head shot away save the vital brain and spinal cord. The author's point is that he, as a surgeon, sees no difference in a scale of horror between this and the effects of chemicals. Ask a contemporary surgeon who has treated high-velocity bullet wounds (one who has served in Northern Ireland will be experienced) about the relative horrors of the two causes of injury, chemicals and missiles. One cannot establish a hierarchy of horror.

Another reason for revulsion against chemical warfare is the feeling that, in common with other modes of modern war, it is impersonal, remote and cowardly. It does not fit the old heroic image of the warriors engaged in man-to-man combat, an image that to a degree can be invoked if you can see and shoot someone directly. Indirect modes of war have always attracted contempt from the front-line warrior, as for example when Suleiman's sappers undermined the fortifications of Rhodes and sealed the fate of the Knights of St. John, or the continuing debate about "carpet bombing" of cities in World War II. On the other hand, the carnage of direct combat has no relative attraction; the film version of Henry V, starring Kenneth

Branagh, gave one of the most realistic and gory reconstructions of such combat in the Agincourt scenes. Perhaps the only thing in favour of the direct action is that the perpetrator sees the effect of his actions; the blood flies into his face.

I see no profit in the discussion of degrees of acceptability of war; all is horrific. The reason for attempting to ban chemical war is to remove one little piece of horror from the whole mosaic. If we are successful in this, then we can go on limiting further and further the scope of war until none is left. This is the prosaic, bureaucratic method of limiting the problem, as opposed to the grand, dramatic gesture of renunciation. The former is, in my judgement, the one most likely to succeed, but only as regards limitation of war. The final removal of war depends on an understanding of why people are aggressive and go to war, as outlined in Chapter 2.

In the late 1970s, the chances of eliminating chemical war were good as few nations had chemical weapons and none seemed ready to use them. However, the Iraqi acquisition of these weapons and their use in war with little international censure has caused other nations to consider the utility of a chemical arsenal. Furthermore, nations with no previous interest at all in chemical war have now to face the alternatives of wholeheartedly supporting an enforceable ban on chemical weapons, or of acquiring the means of defence against them and perhaps an offensive capability. The situation now is that there is a much greater urgency to introduce an international agreement, but at the same time the situation is much more complex than it was in the 1970s.

The technical problems that confront the negotiators of a Chemical Weapons Convention have been discussed previously in different parts of the text, or are implicit in the descriptions of the technology that have been given. They are drawn together in Table 9.1.

The first problem, that of saying whether something is a chemical weapon or not, might seem simple, particularly as it has been discussed in Chapter 5, but on further thought it becomes so complex as to be one of the factors threatening agreement on a convention. Starting with the chemical itself, we can try to define a weapon by a measure of toxicity, of which one value used has been that of a dose equal to or less than 0.5 mg per kg body weight injected subcutaneously to cause death. This value has been used to define supertoxic lethal chemicals. It is purely arbitrary and must be seen simply as a guide to separating very toxic chemicals from others of lesser concern. It fails, however, to include mustard; 35 mg of liquid mustard injected subcutaneously into a man weighing 70 kg will make a horrible mess at the site of injection but is unlikely to kill him. There are also many chemicals that are lethal at very much less than 0.5 mg/kg, yet are

Table 9.1. *Technical problems confronting the negotiators*
of a Chemical Weapons Convention

1. *Definition of what is a chemical weapon*
   Toxicity criteria
   Dual-purpose chemicals, those with civil and military uses
   Intended use of the chemical or device
   Union of toxic chemical with delivery system
   Borderline chemicals; riot-control agents, defoliants, insecticides, etc.

2. *Disposal of existing stocks*
   Safety, especially concerning old and leaking weapons
   Avoidance of pollution

3. *Methods of checking compliance to a convention*
   Inspection of chemical factories and stores
   Inspection of military stocks in magazines

4. *Control of transnational movement of chemicals related to warfare*

5. *Determination of chemical use in war*

unlikely to be used in chemical weapons because they are solids (Chapter 5). Certainly a chemical must be very toxic to be of use in a weapon, but all toxic chemicals are not potential weapons.

If the chemical has no obvious use, the investigator can justifiably enquire what it is for; no one makes chemicals for fun. The reply may be that it has a clear industrial use, as does phosgene for the synthesis of other chemicals, or is an unavoidable by-product that the industry dose not want; hydrogen cyanide sometimes fits this description. Such dual-purpose chemicals need to be considered carefully in any wording of a convention. If the chemical is located in conjunction with a delivery system that is clearly a military munition, then its intended use becomes much easier to define, or more difficult to explain away, depending on your role as investigator or guilty party.

I have mentioned before (Chapter 5) the difficulty of classifying chemicals that are used, or could be used, in warfare, but not as lethal, person-directed weapons. The riot-control agents (tear gases) could be banned for use in war but permitted for domestic riot-control purposes. Herbicides used as defoliants to deny the enemy cover pose a problem to the draft treaty because, when they are used at a military base camp to clear weeds, they are obviously not playing an aggressive role. Insecticides would seem at first to be entirely innocent, but they can cause confusion. An army in the field will have problems with insects, for example, with mosquitoes in humid areas (Vietnam) or flies in the desert (Saudi Arabia). An extensive spraying program will be necessary, and this will lead to some suspicions of

Table 9.2. *Clues to the identification of chemical weapons,
by examining either the chemical or the munition*

The chemical
1. Highly toxic, usually lethal at low dose
2. Readily enters human body (volatile liquid)
3. Produced in quantity (hundreds of tonnes for regular warfare)
4. No obvious civil use

The munition
1. Contains liquid, rather than solid explosive
2. Thin-walled containers, not designed for shrapnel effect or for armour piercing
3. Small quantity of explosive as bursting charge
4. Peculiar or unusual fittings, e.g., filling plug, corrosion resistant lining

Note: Make sure the munition is not a smoke round, or a pyrotechnic device for signalling or other uses.

intent. Why is this nation, which is engaged in a military campaign, importing the precursors to make organophosphorus esters? What are those aeroplanes doing spraying overhead? Our laboratories report residues on samples of leaves and soil that could be the breakdown products of a nerve agent: Is this chemical war? Well, of course it is, but insects are the target, not humans. The herbicide problem will be difficult to solve. The insecticide one does not directly impinge on the draft treaty at all, but the confusion it could cause has to be acknowledged.

No one factor or property suffices to define a chemical weapon; the identification of the weapon must be the result of a matrix of observations leading to one conclusion. Table 9.2 lists some of the properties an investigator would look for in a chemical in a factory or bulk store, or alternatively in a suspect munition. The factors are essentially a recapitulation of the preceding discussion, with some extra points. Production in quantity is a necessity for a chemical intended for a war; chemicals are area-coverage weapons and wasteful. This is a useful indicator to distinguish weapon chemicals from pharmaceuticals, for a country does not need hundreds of tonnes of a highly active drug. The munition properties are also worth study, bearing in mind the objective is to deliver a fair quantity of chemical (not steel) on target. I once had the task of investigating a supposed chemical weapon, which was in fact a standard rocket-propelled grenade (RPG-7) designed to penetrate tank armour. The design (Fig. 9.1) of this rocket is at first odd because of the space or "stand-off" at the front, which one could mistake as a chamber for a chemical fill. The stand-off, in fact, is designed to present the shaped charge of plastic explosive at the right position for maximum

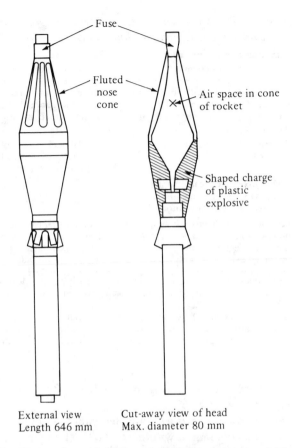

Fuse

Fluted nose cone

Air space in cone of rocket

Shaped charge of plastic explosive

External view
Length 646 mm

Cut-away view of head
Max. diameter 80 mm

Figure 9.1. The PG-7 rocket with high explosive antitank warhead. This is launched from the shoulder-fired RPG-7 launcher. Note the empty stand-off space in the cone of the rocket, which might be mistaken for the space for a chemical fill.

effect, which is like that of an oxy torch rather than just an explosion. The RPG-7 weapons are, of course, the favourite of revolutionaries around the world, along with the AK-47 gun.

The lawyers drafting a convention are going to have a difficult job defining a chemical weapon, for as you will realise, the translation of the concepts just outlined into unambiguous legal language will not be easy. Any one element of the weapon concept (e.g., toxicity criteria alone) is not sufficient as a definition, but a matrix of many elements will be highly successful in recognising a chemical weapon. Legal interrogators like yes or no

Table 9.3. *The three schedules of chemicals proposed
under the Chemical Weapons Convention*

Schedule	Type of chemical	Method of enforcement of controls
1	Known chemical agents, their close structural analogues and their immediate precursors with no legitimate civil use.	Possession limited to very small amounts (national aggregates of 1 tonne) subject to strict inspection and control
2A	Chemicals that are precursors of chemical agents	Annual declarations of stocks will be required, checked by on-site inspectors
2B	Highly toxic chemicals that are not precursors but may threaten the objectives of the convention	
3	Chemicals of concern but which are produced in large quantities by industry for legitimate purposes—e.g., hydrogen cyanide and phosgene	Data relating to production and consumption will be reported, spot checks possible, indications of weaponisation will be looked for.

Notes: Control of Schedule 1 chemicals is by a ban on militarily significant quantities. Schedules 2 and 3 are to be controlled by data reporting backed by varying degrees of active inspection.

The term "dual-purpose chemical" is sometimes used. It could refer to Schedule 2 and 3 compounds that have both civil and military uses, but it is used most often to indicate Schedule 3 chemicals (such as phosgene) that could be used in weapons without chemical modification.

answers, so that it will be interesting to see whether the lawyers can cope with a matrix of partial answers.

The draft convention thus has a strong element of intent in the definition of chemical weapons contained within Article II. The wording (current in a 1991 draft) is: "toxic chemicals . . . and their precursors . . . except such chemicals intended for purposes not prohibited under the Convention, as long as the types and quantities involved are consistent with such purposes." The draft goes on to include specifically designed munitions, devices and associated equipment within the compass of the chemical weapon definition. The aim is to force a challenged party to explain the unusual chemicals, and the quantities held, in accord with my discussion. In addition to an overall definition, the convention will probably list specific chemicals of concern, of which it is proposed there be three schedules, as in Table 9.3. Schedule 2 currently has two parts, A and B, but the final text may well be different.

The problem now moves from one of technical semantics to a purely practical problem of how we get rid of what we have got (Table 9.1).

By "we" I mean a collective, world-citizen "we" taking responsibility for those nations that have chemical weapons, which are the former Soviet Union (40,000 agent tons), the United States (30,000 agent tons), Iraq (unknown quantity, currently being investigated by the United Nations), France (which has declared it has no chemical weapons, but may have agent stocks not weaponised) and presumably several others. The Soviet Union and the United States signed a bilateral agreement in 1990 expressing the intent to reduce their stocks to 5,000 agent tons each by the year 2002. Both parties now face technical difficulties in carrying out this promise. The United States is attempting to destroy its chemical weapons stocks by high-temperature incineration, with the first big exercise in this respect conducted on Johnston Island in the Pacific Ocean in 1990. It has already disposed of minor quantities of weapons by incineration and by hydrolysis in the United States. The Soviet Union made one attempt at destruction by hydrolysis at Chapayevsk in the late 1980s but encountered serious environmental problems with the disposal of the liquid waste. It is generally believed that the United States will have to help them with the technology of destruction, when the uncertainties attendant on the breakup of the Soviet Union are resolved. The Iraqi stockpile will be very difficult to clean up in its half-destroyed state. The UN Special Commission has to decide whether the Iraqis will be entrusted to destroy it, or whether it will be done by international teams. It is certainly hoped that Iraq will pay the costs, expected to be very high (tens of millions of dollars).

The two main methods of destruction are by high-temperature incineration or by chemical hydrolysis. Neither method meets the requirement of one Greenpeace spokesperson, who wanted complete destruction with zero production of residues. The Law of Conservation of Matter gets in the way of this optimistic demand, for what goes in at the front end comes out in equal mass but different chemical form at the other end. Thus destruction is conversion to a different chemical form: Nothing is lost or disappears; the substance merely changes. During incineration the change is from chemical agent and air to the ultimate oxides of all atoms in the agent that can form oxides: CO_2, H_2O, P_2O_5, HF for Sarin, CO_2, H_2O, SO_x, HCl for mustard, also some NO_x from nitrogen in the air. Hydrolysis means breakdown by water, but in practice for nerve agents an alkaline water medium works much faster, to produce initially isopropyl methylphosphonic acid from Sarin, together with a fluoride salt. The breakdown can be pushed further, if desired, by appropriate conditions. Hydrolysis of mustard is less easy because the liquid does not mix with water, and therefore chemical reaction has to be aided by stirring or the use of surface-active agents (detergents). The products are complex; comprising thiodiglycol, hydrochloric

acid or chloride salts and organic sulfonium salts. Hydrolysis, whether of
nerve agents or mustard, produces more complex mixtures than does incineration. The products are entirely liquids, of a volume perhaps six times
that of the agent consumed. The products of incineration are gases and solids, of which the acidic ones can be removed from the flue gases by scrubbing with alkaline water solutions, producing phosphate, sulphate, chloride
and fluoride salts, so that what is discharged to the atmosphere is mainly
carbon dioxide (CO_2) with some SO_x and NO_x.

The two obvious concerns about these destruction processes are, first,
their safety for the operators and others in the neighbourhood and, second,
the possibility of environmental pollution. A destruction system must be
well planned and engineered to contain agent liquid and vapour so that the
operators are not at hazard. The basic problems and solutions can be de-
duced from Chapters 5 and 6; in reality a well-designed plant will employ
remote handling devices to open and process the chemical munitions. Incineration is a predictable method, and the products are, as outlined already, contaminated to very low levels (parts per billion, possibly) with
very stable organic compounds. If the mix being incinerated contains cy-
clic aromatic compounds, such as occur in some plastics, then there is a
greater chance of the formation of environmental pollutants such as polychlorinated (or polyfluorinated) dioxins, furans or biphenyls. This formation of undesirable compounds is by no means inevitable, but the output of
the incinerator needs to be checked for them. The salt solutions resulting
from the scrubbing process may also contain very low levels of complex
pollutants; the United States proposes to bury the solid waste from the
Johnston Island incinerators in the continental United States. It should, in
fact, be an innocuous mixture of salts, tainted not by actual ingredients but
by the mental association with its origin. Some polychlorinated organic compounds and some heavy metals (from metal munition cases) will be present, but at very low concentrations; the potential for pollution is thus low.

Hydrolysis produces a large volume of liquid wastes, which have to be
further disposed of. Unlike the salts from the scrubbers of the incinerators,
this liquor still contains complex organic matter, the carbon–carbon chains
of the agent molecules. As stated before, this will be a complex mix of compounds, which is unlikely to be useful as feedstock for the synthesis of other
compounds, although perhaps it could be if the hydrolysis process were
strictly controlled. Anyway, the liquid product must be handled carefully to
prevent pollution. Hydrolysis does not cause air pollution, apart from that
which may result from the escape of agent or product vapours.

My own opinion is that a well-designed incineration plant will lead to less
pollution and be the most effective way of getting rid of the chemicals,

given the current state of technology. If better processes become evident, however, I am willing to reconsider. Newer processes for breaking down chemicals are being examined but are far from being in the practical engineering stage. It will be quite evident to you that the process of destroying this weaponry is going to be extremely expensive – probably much more than the cost of production.

Our next problem really is a problem, both technically and diplomatically. I do not intend to discuss the latter aspect, other than to point out the difficulties of negotiating a convention that will require the inspection of chemical factories and munition stores in a sovereign country by a multinational team, and will also expose chemical companies to the risk that their commercial secrets may become available to rival companies. The methods of initiating and conducting such inspections have been the subject of extensive, and on occasion explosive, discussion, yet they are central to the securing of a convention that has teeth. The process of "verification" of compliance, odd wording in my opinion, will be an absolute essential for a worthwhile treaty.

Let us revert to technicalities and consider what has to be done. The scenario is that a team of inspectors, drawn from various nations but now servants of a central body created under the convention, is required to visit a chemical factory or a military storage area to check that no activity relating to chemical weapons is occurring. The team has to arrive quickly with little warning (say forty-eight hours maximum) so that there is not time to clear away incriminating material. It must also be as quick as possible on site so that the normal functions of the factory or whatever are not unduly interrupted. If chemical samples are to be analysed off the site, then arrangements for rapid transport and analysis have to be made. It is obviously preferable that the inspecting team has portable analytical equipment that can do the analyses on site.

First, the factory inspection presents few technical problems apart from the need to avoid being too intrusive and thus being accused of prying into commercial secrets. There is no technical problem in taking samples from all parts of the process line and from all chemical stores, doing complete chemical analyses and deducing that all is well. It is also possible to be certain of finding traces of suspicious chemicals that show the plant has rapidly switched from one process to another in the last twenty-four hours. Such a complete survey may or may not be acceptable under a convention, but industry certainly will not be happy about it. The question then arises as to whether the analytical technology can be developed to show that certain chemicals are not there, without accumulating a complete catalogue of chemicals that are there. After all, it is absence that we need to check on and confirm. It seems at present that such a concept of inspection is technically feasible, but it has not yet been worked out sufficiently to enable us to be totally positive about it. Various items of chemical analysis equipment can be tuned to respond only to certain chemicals or to particular families of compounds, so that a negative response at the factory with a detector tuned to nerve agents is sufficient to allay suspicions that nerve agents are being made there. If a positive response is obtained at a factory claimed to be making insecticides, however, then further analysis is needed because some insecticides have a structure similar to that of nerve agents. (They are both organophosphorus esters.) If the factory was stated to be making household detergents, then a positive response would be indicative of a very suspicious situation.

The chemical agent monitor (CAM) is one example of a portable detector that can be tuned to detect groups of chemicals, and the gas-detector tubes of the Drager type are simple but useful tools. These are described in Chapter 6. More complex analytical equipment such as mobile mass spectrometers can be modified to restrict the information output to that required to be known under the terms of the inspection. One way to do this would be to modify the software that processes the signals from the actual instrument and transforms them into output data as a graphical or numerical record. The software package to be used could be agreed to by the inspectors and the industry, so that only relevant data are available to the former.

It seems to me that there are no insoluble technical problems in developing an inspection regime for industry but that complex procedural matters will need sorting out. These are now being addressed in a number of countries, which are conducting trial inspections of chemical factories. Perhaps the main aim of such inspections should be to reassure industry in general that these inspections will not compromise their commercial secrets.

The inspection of a military storage area needs a slightly different approach. The team will be confronted with a series of packages, artillery rounds, rockets, aircraft bombs, mines or whatever, and will be required to check that these packages do not contain toxic chemicals, as opposed to high explosive, smokes, pyrotechnic devices or other accepted contents. In a short inspection the team cannot open the munitions, which anyway cannot be done in situ in the magazine, due to the toxic nature of the possible contents. The external appearance of some munitions may be enough to allay suspicion; thus the absence of unusual features such as filling plugs or discrete burster charges will be reassuring. In other cases, it will be necessary to try to discover what is inside the munition, through 10 or 20 mm of steel case. Portable x-ray or ultrasonic equipment can probably be developed that will give information on the physical geometry of the contents, that is, distinguish the solid, shaped charge of an armour-piercing round from a mobile liquid fill with a distinguishable meniscus. Chemical analysis through a steel wall is more of a challenge, but in recent years a number of techniques have been developed in which some form of radiation is used to bombard a sample, and the secondary radiation emitted from that is captured and examined. This secondary radiation can give information on the chemical structure of the sample. Furthermore, the radiation used in some techniques is capable of penetrating steel. It is probable, therefore, that portable equipment will be developed that can allow a quick but thorough examination of a magazine.

Some situations will cause problems. For example, if the inspection team finds rocket cases in a magazine, complete with propellant charges but no contents, what does it do? It is very easy to store toxic chemicals elsewhere in cylinders of resistant plastic, ready to be slipped into the rocket case and fired. To meet such circumstances, the inspectors must have the right to be answered on all questions relating to unclear use of munitions.

Item 4 of our list of problems (Table 9.1) is the control of the transnational movement of chemicals related to chemical warfare. In Chapter 8, I discussed the problem sufficiently to illustrate the complexity of the task, but admit to a paucity in solutions. The reader will see that controls can be evaded, but probably not for long or on a scale that would allow the production of a large chemical arsenal.

Our last item is the determination of whether chemicals have been used in a conflict. I have read some extraordinary comments on this. In one article it was written that (1) the use of chemicals against an unsuspecting foe is extraordinarily effective and militarily decisive and (2) it is very difficult to determine whether chemical weapons have been used a few days after the event.

If the weapons are so effective, the evidence is all around. What are these corpses with no bullet holes or shrapnel wounds? Why has this soldier lost the skin from his arm when he says there was no fire or explosion? Why are the pupils of this man's eyes constricted to pinpoints? A militarily effective concentration of chemical agent will leave chemical residue around in high concentrations for days (even GB will not evaporate totally if it has soaked into earth or vegetation). Any flat surface, a few square centimetres, is all the analyst needs to make a positive identification. The sample collector should ignore any visible deposit but systematically sample flat surfaces at various distances from the centre of the event. Ignore any yellow spots, unless you have an intense interest in the diet of flower-feeding insects.

Table 9.4. *A comparison of two cases of the suspected usage of chemical warfare agents, showing factors that varied widely in the two cases*

Factor	First case, Southeast Asia	Second case, Iran
Origin of samples	Some doubt	Certain
Chemical suspected	No trace of known chemical warfare agent	Conventional chemical warfare agent
Analysis required	Unusual (mycotoxins were suspected)	Well known
Quantity of sample available	Near limit of detection	Plenty
Quality of conclusions to be made	Inconclusive	Definite

Note: The comparison relates to a scenario in which samples from an area of suspected chemical attack are brought to a laboratory for analysis.

If it is difficult to determine whether a chemical attack has occurred, the most common reason is because an attack did not occur. There is great propaganda value in saying that your opponent is an inhuman user of chemical weapons; many of the reported incidents of chemical usage since World War II are undoubtedly false. There was no doubt of chemical usage in the Gulf War by Iraq; the casualties and the munitions were all too evident (Table 9.4).

Therefore in technical terms there is not the slightest problem in determining whether chemicals have been used on a battlefield after four or five days, provided a sensible collection of samples is made, and supposed casualties are examined by a medical officer experienced in the endemic diseases of the region. Some of the reports[2,3] on the investigation of the so-called Yellow Rain in Laos and Kampuchea, the suspected use of mycotoxins as warfare agents, illuminate this topic, or rather demonstrate how muddled analysis of doubtful data can lead investigators astray.

My conclusion is that there is no insuperable technical reason that would deny the implementation of a Chemical Weapons Convention. Certain technologies need to be developed further (e.g., on-the-spot analysis, examination of munition contents) but all problems are technically soluble. The real problems are the diplomatic, political and military ones, which will keep the negotiators sparring round one another for a while yet. In other words, war is a human problem that requires human solutions; we have the technical answers.

References

1. Vedder, E. B. (1925). *The Medical Aspects Of Chemical Warfare.* Baltimore: Williams and Wilkins.
2. Evans, G. (1983). *The Yellow Rainmakers.* London: Verso Editions.
3. Harris, E. D. (1987). "Sverdlovsk and Yellow Rain." *International Security.* 11:41–95.

10

The future of arms control

There is a long history of attempts to control the horrors of war by international agreement, so that the negotiators of a Chemical Weapons Convention have plenty of good examples to profit by and bad examples to avoid. These past agreements divide into the more general ones on topics such as the minimum treatment to be accorded to prisoners of war, and the specific ones relating to a particular type of weapon.

The treatment of prisoners of war has been increasingly clearly defined in a series of agreements commencing with the 1874 Brussels Declaration (which was not ratified and did not enter into force) through the Hague Conferences of 1899 and 1907, the 1929 Geneva Convention and that of 1949. What was the value of this effort, for example, in World War II? The treatment of prisoners varied markedly, largely in accord with the ethics of the capturing military, and also varying with the intensity of the fighting. We are told that the Japanese maltreatment of prisoners was due to the fact that the Japanese military ethic did not allow of a person surrendering, and therefore opponents who did so were nonpersons in the military view. Allied soldiers captured by the Wehrmacht received correct if harsh treatment, whereas those trapped by the Waffen SS might well be machine-gunned. I have no doubt that the Allied treatment of prisoners was also imperfect at times, owing to casual brutality, indifferent neglect and sometimes systematic revenge. Nevertheless, the 1929 Geneva Convention gave some excuse for intervention, exerted some moral pressure and probably helped many prisoners. The real value of such conventions lies in the fact that an agreed standard does exist, and there is a yardstick against which your own or the enemy's conduct can be measured. Therefore, although the agreement may not prevent the performance from falling below standard, at least this failure can be quantified and eventually publicised. A standard can also be useful in war crimes trials, although I am rather cynical about these; they are a case more of *vae victis* than of justice. However, the 1946 Nuremburg war crimes trials at least served to record formally some of the Nazi misdeeds.

The regulation of specific instruments of war commenced with the 1868 St. Petersburg Declaration on explosive projectiles under 400 g weight, continued with the 1899 Hague Declaration on expanding (Dum-Dum) bullets, and then went on spasmodically with various topics to the 1981 UN Convention on excessively injurious or indiscriminate conventional weapons. At the same time, of course, there were the measures to control chemical warfare that we have traced out in Chapter 3.

The topic of the explosive/expanding and high-velocity bullet affords a long and instructive history, which we can examine as an illustrative example of technical problems in arms control. The 1868 St. Petersburg Declaration was readily agreed to as a means of reducing unnecessary wounding from small arms: "and that Commission, having by common agreement fixed the technical limits at which the necessities of war ought to yield to the requirements of humanity..." The technical limits here referred to were the exclusion of explosive, fulminating or inflammable substances; the 400-g maximum is a separate consideration designed to allow the use of explosive artillery shells to continue. The 1899 declaration on expanding bullets prohibited the "use of bullets which expand or flatten easily in the human body, such as bullets with a hard envelope which does not entirely cover the core or is pierced with incisions." The narrow technical meaning of the 1899 declaration is clear enough, but considered in the broader sense of the intent of both the 1868 and 1899 declarations, there is a technical controversy that has continued ever since. Large, jagged and diffuse wounds can be produced by bullets other than those which explode or expand. For example, a bullet that has a tendency to tumble end over end when it hits flesh will cause a large wound. Such effects may occur with high-velocity bullets, or with certain designs of bullet and gun. Tumbling bullets may be regarded as against the intent of the 1868 and 1899 declarations, but are not clearly proscribed by the technical wording. This problem surfaced again in the 1970s. Armies in World War II largely used the rifle calibres of 0.303 inch (the Lee-Enfield rifle) or 9 mm (many German weapons). In the 1950s the new self-loading rifles used the old 0.303-inch calibre, now in its metric equivalent of 7.62 mm, but these were still relatively heavy, and also a load of ammunition was weighty. There was increasing interest in smaller calibre rifles, of which the Colt Armalite at 5.56 mm was a favourite example.

Now, the momentum of a missile is the product of its mass and its velocity. If you reduce the mass, you need to increase the velocity to get the same (or nearly the same) effect upon impact on the target. The 5.56-mm calibre bullets were therefore given a greater velocity, but this does not give exactly the same terminal effects. A low-velocity, heavy bullet crashes along its way, is not greatly deflected by intervening obstructions and expends its energy on the target as a solid thump. By contrast, the high-velocity, light bullet deflects off intervening surfaces, tends to tumble and instantly transfers its energy to the target. The effect is that of a shock wave, rather than a thump. The surgeon sees a small, neat hole in the front of the victim's body, but at the rear there is no exit hole, but rather a large area of flesh blown out of the back. The wound is a cone, the apex at the front. This may be further complicated if the projectile has tumbled, destroying more living tissue as it does so. These wounds are difficult to clean up, as the area of damage to the tissues extends beyond the zone of obvious physical disruption, with the consequence that the surgeon has to guess how much debris to remove. Because of these mutilations caused by high-velocity bullets, there was much discussion and experiment in the 1970s on the possibility of defining a maximum velocity for small-arms bullets that would be permissible under international law. However, this question was not settled, and the UN convention on excessively injurious or indiscriminate conventional weapons of 1981 did not include the topic, although it had been discussed at the two conferences that determined the text of the convention.

Why was this control measure not accepted? There seem to have been two main reasons, the first being that it is technically difficult to define a maximum velocity as the unwanted effects increase steadily with velocity; there is no magic transition point. Second, by 1980 many nations had in production and service high-velocity, 5.56-mm calibre weapons they were not prepared to scrap. This failure may well be a portent of doom for the Chemical Weapons Convention; failure due to difficulties of technical definition and to the intractability of possessor nations is not unlikely.

An ostensible success in the control of arms internationally is the Nuclear Non-Proliferation Treaty, signed in 1968 and subsequently attracting a membership of 141 states (by 1990). The treaty seeks to prevent the proliferation of nuclear weaponry beyond those five states (the United States, the former Soviet Union, the United Kingdom, France, China) recognised as possessing these weapons in 1968. It has provisions to prohibit the movement of nuclear weapons, or materials for their construction, between nations acting either as donors or recipients. This control is supervised by the International Atomic Energy Agency. The treaty also has obligations on states who are members to assist in the development and dissemination of peaceful uses of atomic energy, and to pursue measures aimed at stopping the nuclear arms race and obtaining complete nuclear disarmament (Article VI).

There is some justification for regarding the Non-Proliferation Treaty as the most successful of arms control treaties. Thus it has reduced the haphazard spread of nuclear weaponry and demonstrated that technical controls can be devised to limit particular forms of weaponry. It therefore serves as a model for a Chemical Weapons Convention, but a model that has demonstrable imperfections.

The Non-Proliferation Treaty has a number of features that many states and individuals see as major faults. First, it has frozen the states as of 1968; that is, possessor states remain in possession of nuclear arms. Many small, developing countries see this as the maintenance of military mastery by domineering rich countries, using the treaty as a device to keep the underprivileged nations in a subservient state. This feeling would be tempered if the developing nations could see any real attempt to implement Article VI and secure nuclear disarmament rather than nuclear freeze. Unfortunately, the possessor nations have made little (or no) progress here, and the article has no stated program for disarmament to occur.

Second, several important states have not acceded to the treaty, including two possessor nations, China and France. The latter has continued regular nuclear weapons testing in the Pacific, much to the annoyance of neighbouring countries. Other nontreaty nations include some suspected of developing nuclear weapons programs, such as India, Pakistan, South Africa, Israel and Brazil. These few nonsignatories are obviously more significant than many of the minor signatories such as Cape Verde, Bhutan, Fiji, Tuvalu, Seychelles, to name just a few. The number of treaty adherents is not a good guide to its success, for as long as some critical nations remain outside, it will not be seen as achieving the desired aim of total nuclear arms control.

The lessons for a Chemical Weapons Convention are clearly as follows:

1. A nonproliferation treaty will not satisfy developing or Third World countries. The convention will have to include rigid agreements that chemical weapons stocks are destroyed by all participants.

2. It will have to be seen as primarily a disarmament treaty, in which all signatories are equal, and no special privileges attach to possession of chemical weapons at the time of signing. There can be no element of patronage or superiority.

3. The original signatories will need to include all nations known to possess chemical weapons and all nations with major chemical industries. Intractable countries (e.g., Iraq) that clearly possess such weapons thus pose a major difficulty in concluding a convention.

The Non-Proliferation Treaty will therefore cast a cautionary shadow on any Chemical Weapons Convention. Its success in a technical sense (i.e., the international implementation of safeguards) may be seen as a positive encouragement to a chemical treaty, but a degree of caution is necessary here. Nuclear fuels are well characterised, have few places of origin and require special handling. By contrast, toxic chemicals are of many different types, can be produced in many factories or plants and can be handled by simple methods (with some degree of risk). Therefore the success of nuclear proliferation safeguards may not be a true indication of the likely success of technical methods to control chemical weapons.

Overall, the Non-Proliferation Treaty is of value to us in considering the control of chemicals, provided we make sure we are learning the right lessons from it. We can in haste get the wrong lessons.

None of the international agreements mentioned previously in this chapter is a disarmament agreement; they are measures of control or limitation but do not require the rejection of a whole class of weapons. There are aspects of disarmament in them, such as limitations on some types of conventional weapons in the 1981 convention, and at least an intention within Article VI of the Nuclear Non-Proliferation Treaty. Only one true disarmament measure has been internationally agreed, and that is the Biological and Toxin Weapons Convention of 1972. This bans the development, production and stockpiling of biological material that could be used in war, or has no use other than an offensive military one. This convention quite obviously is a model that must influence the embryonic Chemical Weapons Convention; to some people it is, however, a bad model from which the lesson to be learnt is how not to do it.

The convention was concluded fairly quickly with support from the United States as the result of a personal interest of President Nixon, who had also imposed a moratorium upon the production of new chemical weapons. Ironically, the United States has been the chief critic of the convention ever since, for after a short honeymoon period in the 1970s, that country accused the Soviet Union of breaking the convention, and criticised the weakness of the convention that allows it to be flouted. In the late 1970s and 1980s the U.S. government repeatedly charged the Soviet Union with breaking the convention by the development and weaponization of toxins and infectious agents. Countercharges against the U.S. government came from Cuba (among other nations) and from within the United States, but the Soviet Union confined itself to accusations that the United States employed chemical warfare in Vietnam. It is my intent not to judge or weigh these charges but to find the technical weakness in the 1972 Biological Weapons Convention.

That weakness is not too hard to find, for the convention is devoid of any provisions to cope with a nation that breaks the convention—in fact, it is deficient in two degrees: First, there is no regular mechanism to ensure that countries comply; there is no inspection or challenge procedure that can be activated when suspicions are aroused. Second, if suspicions become strong, the only form of censure is by reference to the UN Security Council, a move that can be blocked through a veto by a permanent member of that council.

The kind of situation that can arise is well illustrated by the Sverdlovsk incident, of which a clear account is that of Harris.[1]

A series of reports appeared in Western newspapers in 1979 relating to an accident at a Soviet bacteriological warfare installation. These accounts contained vague and slightly conflicting details, but implicated a military installation at Sverdlovsk, from which bacterial contamination had been accidentally released to cause illness and deaths in a nearby village, Kashino. No official statement was made by the Soviet government, until the United States privately asked for an explanation in March 1980, and then the following day made a public statement, which could be understood to accuse the Soviets of breaking the 1972 convention. This brought a swift denial from the Soviets, who said that an outbreak of anthrax had occurred, but it was due to the spread of anthrax-infected meat from black-market cattle slaughtered in a backyard and distributed in the neighbourhood. There is, after ten years, still no way that independent observers can determine the truth of what happened. What is clearly revealed is the weakness of the convention.

The United States had suspicions that the convention had been breached, whether on good or bad evidence is irrelevant. The Soviet Union made no move to meet these concerns, other than to issue vigorous denials. Presumably because the United States perceived that its evidence was weak, it made no formal challenge within the terms of the convention. Yet the Soviet Union was obliged to "consult and cooperate" in the situation, which it did not do. Nor did it do the obvious thing that would allay suspicion: bring in a team of scientists and medical persons from some neutral countries to define exactly what had happened.

Had such a team found that a Soviet laboratory had been working on anthrax cultures, it would come up against another weakness of the convention: How do you tell whether the material is for "prophylactic, protective or other peaceful purposes" or for war?

Certainly, people may have differing opinions about the accusations and counterclaims on biological warfare activities that have passed from West to East and back, but I think everyone must see the deficiencies of a legal agreement that has no provisions to ensure that it is complied with, and the importance for us lies in the absence of scientific methods to ensure compliance. The Biological Weapons Convention is subject to periodic review conferences (one was held in 1991), which are endeavouring to bolster confidence in the convention by increasing openness between participants. This may eventually succeed, but is a retrospective way of improving a measure deficient at its origin.

The almost universal poor regard for the Biological Weapons Convention may be sufficient to bring all similar agreements in disrepute; it will undoubtedly ensure that no Chemical Weapons Convention will be signed unless it can demonstrate that it has procedural and technical provisions that will give a high reassurance of compliance.

I entitled this chapter "The Future of Arms Control," then talk about past failures. My excuse is that the past will strongly influence the future, and you should be able to make your own assessment of the future of arms control from these observations.

Reference

1. Harris, E. D. (1987). "Sverdlovsk and Yellow Rain." *International Security* 11:41–95.
   General information on arms control agreements can be found in
   A. Roberts and R. Guelff (eds.) (1982). *Documents on the Laws of War.* Oxford: Clarendon Press.

11

◁══════════════════════════════▷

Conclusions; or, Problems defined

Conclusions are uninviting diet for the reader: With the juice and meat gone, the bones and gristle do not attract; yet it is worthwhile reviewing the framework of an argument or discussion so that the structure may be further augmented in consent or torn down in angry objection. I want to put my summary and conclusions into two sections, the general background information and then the technical arguments specific to a Chemical Weapons Convention (Table 11.1).

I have tried to show that the abolition of aggression and war has a scientific aspect, for we need objective knowledge of what makes the human animal aggressive before we can try to mould society in such a way as to remove those identified causes. Behavioural science must therefore blend into the humanistic studies of man to produce a world in which war is unknown. At present we are far from understanding the basic causes of aggression, let along remedying them; so our best course of action is to limit warfare in a piecemeal manner. Chemical warfare is one expression of war that seems susceptible to limitation.

Chemical warfare has a comparatively short history, although its antecedents are long. Its extensive use in 1915–18 did not appear to make any decisive change in the long stalemate on the western front; in fact, the only conflict in which it may have influenced the outcome was the Iran–Iraq conflict of the 1980s. However, the technology of this kind of warfare is more mature; so both the toxic chemicals and the munitions to carry them are efficient instruments of war. Chemical weapons are better suited to the modern era of missiles and rocketry than they were to the former times of the artillery gun.

Defence against chemical weapons has also developed technically to the point where well-equipped and trained soldiers can be protected against

Table 11.1. *Chemical disarmament
issues discussed in this book*

General background matters
Science of aggression and war
Chemical warfare, offence and defence
Modern chemical industry
Analytical capabilities: the ability to detect and monitor chemicals
Matters specific to a Chemical Weapons Convention
Definitions of chemical weapons and related items
Destruction of existing chemical weapon stocks
Technical determination of compliance with a convention
Investigation of alleged use of chemical weapons

chemicals to a degree that would be impossible to achieve against conventional weapons. By contrast, unprotected soldiers are extremely vulnerable to chemical attack; so the effectiveness of this weapon depends very much on the state of the target personnel at the instant when the chemical is delivered. Defensive measures are burdensome to the individual and costly in time for the defending army; hence, to put the matter in ten words, chemical defence is highly effective but very difficult to achieve.

The products of the chemical industry are an integral part of our material life, and the system of production, manufacture and distribution that achieves this is worldwide, not at all restricted by national boundaries.

Some chemical companies have financial dealings that exceed the budgets of many independent nations, and regard themselves as supranational organisations. As a consequence, loyalty is given first to the company and only second to the nation to which the employee happens to belong, or in which the company is sited. The industry is also highly competitive and trade secrets are important, for if the efficiency of production of a chemical can be improved by 5 percent, for example, then this could be sufficient to force a competitor out of the market. Any control of chemicals for use in weapons has to be tailored to work against this background of a ubiquitous and competitive industry.

Since 1970, and more particularly in the 1980s, our ability to detect and identify chemicals has improved to a degree that the older ones among us scientists find extraordinary. It is now possible to detect one chemical when it is present among other chemicals at a level of one part per billion (one in a thousand million). It is also possible to identify a chemical under favourable circumstances when the quantity available is less than one-millionth of a gram. Sensitive chemical analysis is thus available to help enforce a ban on chemical weapons, but it will also make industrial chemists uneasy if they feel their secrets are being discovered; so methods of inspection that are the least intrusive are being investigated.

If we now summarise our discussions on the matters specific to a Chemical Weapons Convention, the first is agreement on definitions. It is quite permissible for a tentative convention to have definitions that differ to a minor degree from those used by other persons or for other purposes. Thus the definition of a chemical weapon in a convention need not accord completely with a military definition, but obviously there must be a basic commonality. The definition must be adequate for its purpose, and as clear and unambiguous as possible to avoid legal distortions of the intent. As we explained in Chapters 5 and 9, the definition of a chemical weapon includes elements of toxicity, battlefield utility, delivery system and intent to use for war. It is, in my opinion, necessary to use both a functional definition and specific examples, taking care that the examples are recognised as such, not as the total listing. The draft convention seems to be going this way, but as the list of examples in the various schedules grows, we must be careful to ensure that this action is not seen as acceptance of each chemical as implicated in war, with the implication that others are innocent. Other definition problems arise with riot-control agents, herbicides and the components of binary munitions. The resolution of these problems will be largely political rather than technical.

My method of defining chemical weapons in general terms, and showing the relationship with other chemicals that might be used in war, is illustrated in Figure 11.1. Many chemicals are used for reasons other than to attack personnel; of those that are aimed at people, the riot-control agents have transient effects only (Chapter 5), whereas the true chemical weapons kill or permanently injure. Note that a convention could specifically ban the use of tear gases in war but permit their use for riot control. Chemical weapons are composed of a delivery device and a chemical warfare agent (the toxic fill), each of which is susceptible to control. I am puzzled each time a chemical by itself is referred to as a weapon, for the delivery device is an essential component.

Destruction of chemical warfare stocks is a complex technical matter but one that can be undertaken safely and without endangering the environment, using existing technology. Certainly we can expect further improvements in technology that will reduce the risks associated with this process, but the necessary technology does exist. Whether it will be properly applied is a different matter.

The function of ascertaining that any future Chemical Weapons Convention is being complied with is a diplomatic and legal matter, but the tools to be utilised will be largely scientific. Recent developments in analytical

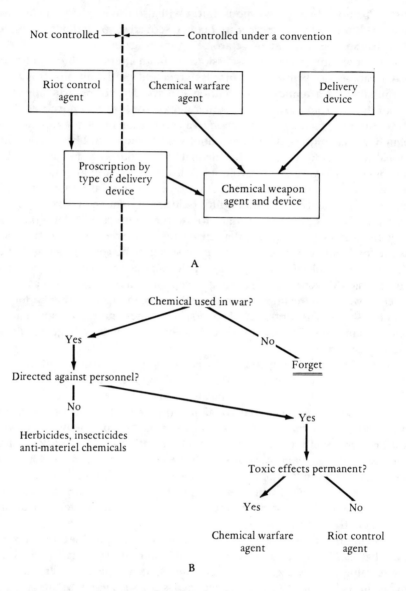

Figure 11.1. Aids to the definition of a chemical weapon. A. A chemical weapon is a combination of a warfare agent (the chemical) and a delivery device. Riot-control agents in particular delivery devices could come within the ambit of a convention. B. The relationship between chemical warfare agents and other chemicals used in war is clarified by a logic tree such as is shown.

technology mean that these tools will probably be effective, although much work remains to be done in developing analytical and chemical monitoring equipment for this particular application. Methods of monitoring industry will need to be developed that do not raise distrust in the chemical industry; sufficient but not excessive knowledge is the requirement. I believe that the appropriate technical tools will be available for the international inspectorate, as it conducts inspections of factories and military stores and alerts the customs services of various countries to the existence of suspect cargos of chemicals.

The technical tools also exist to investigate the alleged use of chemicals in war. This is indeed easy if access to the site is available within days. As the inspecting team becomes more and more remote from the suspect event in both time and place (as with the Kurdish refugees in Turkey 1988), then their task increases steadily in difficulty. The Kurds complained of chemical attack by Iraqi government forces inside Iraq which drove them from the country into Turkey; tens of kilometres and weeks of time separated the investigating teams from the event. Basically, however, there is no technical problem.

The reader must realise that this book has been concerned with the scientific aspects only of the negotiations towards a Chemical Weapons Convention. Many of the difficulties are diplomatic, political and legal matters that I may have mentioned but not enlarged upon. This present work is, therefore, an exposition of a limited aspect of the whole problem and can be regarded as the words of a scientist speaking to the diplomats, politicians and lawyers, as well as to others in general society sufficiently concerned to study this issue. In a wider sense, it is also a study of the function of science in society, through one specific example that requires input from many sources.

In our present Western culture, we have managed somehow to separate parts of that culture into near-watertight compartments, so that the study of the natural world is regarded as entirely distinct from the study of ourselves as humans and creative beings, and from the development and execution of national policy. This is a comparatively recent phenomenon; the sixteenth century, for instance, made no distinctions, as witness Thomas More (public servant, classical scholar and patron of learning) or Thomas Wyatt (ambassador, poet and lover of Caesar's hind). Nor did Islam in its cultural flowerings distinguish between the sciences and the arts. As we are heritors of this recent division of human achievement, we must endeavour to close the breach or at the least act as messengers between science and humanities, and in this case to unite all professions of our society in the strangulation of war.

Appendix 1:
The Seville Statement on Violence

The following statement was issued after a conference at Seville, Spain, of scientists concerned about violence between individuals, groups and nations. It was signed by twenty prominent scientists and issued in an attempt to dispel the notion that war and violence are inevitable parts of human social activity. It was signed in what was, although now totally forgotten, the International Year of Peace, 1986.

Believing that it is our responsibility to address from our particular disciplines the most dangerous and destructive activities of our species, violence and war; recognizing that science is a human cultural product which cannot be definitive or all-encompassing; and gratefully acknowledging the support of the authorities of Seville and representatives of the Spanish UNESCO; we, the undersigned scholars from around the world and from relevant sciences, have met and arrived at the following Statement on Violence. In it, we challenge a number of alleged biological findings that have been used, even by some in our disciplines, to justify violence and war. Because the alleged findings have contributed to an atmosphere of pessimism in our time, we submit that the open, considered rejection of these misstatements can contribute significantly to the International Year of Peace.

Misuse of scientific theories and data to justify violence and war is not new but has been made since the advent of modern science. For example, the theory of evolution has been used to justify not only war, but also genocide, colonialism, and suppression of the weak.

We state our position in the form of five propositions. We are aware that there are many other issues about violence and war that could be fruitfully addressed from the standpoint of our disciplines, but we restrict ourselves here to what we consider a most important first step.

It is scientifically incorrect to say that we have inherited a tendency to make war from our animal ancestors. Although fighting occurs widely throughout animal species, only a few cases of destructive intra-species fighting between organized groups have ever been reported among naturally living species, and none of these involve the use of tools designed to be weapons. Normal predatory feeding upon other species cannot be equated with intra-species violence. Warfare is a peculiarly human phenomenon and does not occur in other animals.

The fact that warfare has changed so rapidly over time indicates that it is a product of culture. Its biological connection is primarily through language which makes possible the coordination of groups, the transmission of technology, and the use of tools. War is biologically possible, but it is not inevitable, as evidenced by its variation in occurrence and nature over time and space. There are cultures which have not engaged in war for centuries, and there are cultures which have engaged in war frequently at some times and not at others.

It is scientifically incorrect to say that war or any other violent behavior is genetically programmed into our human nature. While genes are involved at all levels of nervous system function, they provide a developmental potential that can be actualized only in conjunction with the ecological and social environment. While individuals vary in their predispositions to be affected by their experience, it is the interaction between their genetic endowment and conditions of nurturance, that determines their personalities. Except for rare pathologies, the genes do not produce individuals necessarily predisposed to violence. Neither do they determine the opposite. While genes are co-involved in establishing our behavioral capacities, they do not by themselves specify the outcome.

It is scientifically incorrect to say that in the course of human evolution there has been a selection for aggressive behavior more than for other kinds of behavior. In all well-studied species, status within the group is achieved by the ability to cooperate and to fulfil social functions relevant to the structure of that group. 'Dominance' involves social bondings and affiliations; it is not simply a matter of the possession and use of superior physical power, although it does involve aggressive behaviors. Where genetic selection for aggressive behavior has been artificially instituted in animals, it has rapidly succeeded in producing hyper-aggressive individuals; this indicates that aggression was not maximally selected under natural conditions. When such experimentally-created hyper-aggressive animals are present in a social group, they either disrupt its social structure or are driven out. Violence is neither in our evolutionary legacy nor in our genes.

It is scientifically incorrect to say that humans have a 'violent brain'. While we do have the neural apparatus to act violently, it is not automatically activated by internal or external stimuli. Like higher primates and unlike other animals, our higher neural processes filter such stimuli before they can be acted upon. How we act is shaped by how we have been conditioned and socialized. There is nothing in our neurophysiology that compels us to react violently.

It is scientifically incorrect to say that war is caused by 'instinct' or any single motivation. The emergence of modern warfare has been a journey from the primacy of emotional and motivational factors, sometimes called 'instincts', to the primacy of cognitive factors. Modern war involves institutional use of personal characteristics such as obedience, suggestibility, and idealism, social skills such as language, and rational considerations such as cost-calculation, planning, and information processing. The technology of modern war has exaggerated traits associated with violence both in the training of actual combatants and in the preparation of support for war in the general population. As a result of this exaggeration, such traits are often mistaken to be the causes rather than the consequences of the process.

We conclude that biology does not condemn humanity to war, and that humanity can be freed from the bondage of biological pessimism and empowered with confidence to undertake the transformative tasks needed in this International Year of Peace and in the years to come. Although these tasks are mainly institutional and collective, they also rest upon the consciousness of individual participants for whom pessimism and optimism are crucial factors. Just as 'wars begin in the minds of men', peace also begins in our minds. The same species who invented war is capable of inventing peace. The responsibility lies with each of us.

Seville, 16 May 1986

Appendix 2:
A guide to the terminology of chemical warfare agents

A large number of chemical compounds have been examined for their utility as chemical weapon fills. A much smaller number have actually been put into weapons or used in war. These have been given code letters in a partly systematic manner, and these are given here with a brief description of the type of chemical and its effects. Details of the chemical structures can be found in other books. [1,2] The compounds have usually been given a two-letter code, of which the first letter indicates the general class.

Code letter	Class of compound and examples
H	Blistering (vesicant) agents of the sulphur or nitrogen mustard types. H by itself usually signifies crude sulphur mustard. HD: distilled sulphur mustard. HT: sulphur mustard with 40 percent of T. (See under T.) HL: equal mixture of mustard and Lewisite. HN: nitrogen mustard, of which there are three forms, HN-1, HN-2, HN-3. HS: World War I designation for sulphur mustard, believed to derive from "Hot Stuff." Mustard is also known as Lost, Gelbkreuz (both German), Yperite (French), Y and Yellow Cross.
L	Lewisites, vesicant agents containing arsenic. There are three related compounds, L-1, L-2 and L-3, of which L-1 is that usually referred to. Lewisite was used as an equal mixture with H to give HL.
T	A sulphur mustard more complex than H. In World War II the British used T with H to give HT.
AC	Hydrogen cyanide or prussic acid gas. Kills quickly by stopping the use of oxygen by the body tissues. Cyanogen chloride (CK) can also liberate cyanide in the body, but CK has a choking effect in addition.
C	A designation for choking gases. The C chemicals, however, subdivide into those that damage the lung and can result in death, such as phosgene, and those that irritate and therefore are used as riot-control agents (tear gases).

117

Code letter	Class of compound and examples
C (choking)	Phosgene CG, phosgene oxime CX, cyanogen chloride CK (see also under AC).
C (irritant)	The riot-control agents, of which CS is the safest and most widely used now. CR is a persistent irritant chemical. CN or CAP is still used but has greater longer term toxicity than CS. BBC (bromobenzyl cyanide) was a World War II irritant. See also DM.
D	A range of arsenical irritant materials, with a moderate degree of long-term toxicity. DM or Adamsite was employed as a riot-control agent but is no longer used.
PS	Chloropicrin. The vapour is very toxic. It is still employed as a fumigant in rabbit burrows, but human poisonings have occurred. Not now considered as a warfare agent.
G	The compounds above all antedate World War II. The G agents are the organophosphorus anticholinesterase compounds developed by the Germans in World War II, the nerve agents. The British late in the war had information on "German Gas," hence presumably the G code. The compounds most likely to be used are GA (Tabun), GB (Sarin), GD (Soman) and GF.
V	A series of less volatile nerve agents developed after World War II. VX has been weaponised, and is a persistent and highly toxic liquid.
(V)	V was used in World War II by the British to signify viscous or thickened agent. HT (V) therefore signifies 60 percent H, 40 percent T thickened by chlorinated rubber or perspex.
BZ	A psychoactive compound, that is, one that interferes with signal processing in the central nervous system to produce hallucinations or other disconnection from reality. Weaponised by the United States, but the stocks have been destroyed.

References

1. Clarke, R. (1968). *We All Fall Down: The Prospect of Biological and Chemical Warfare.* London: Allen Lane.
2. Spiers, E. M. (1986). *Chemical Warfare.* London: Macmillan.

Index

Page numbers in *italics* refer to tables and figures.